Obverse Cohainscape Biparic Essentiae and Ellipsoid Conglobations

Odor Terraeus

Obverse Cohainscape Biparic Essentiae and Ellipsoid Conglobations

First Edition September 2025

Written by Odor Terraeus

Contents

Chapter 1

Cohomology of algebraic stacks – a review

In this chapter, we collect some results on cohomology of stacks, all of which are well known. We will need all these results in Chapter 2 and Chapter 3.

In the first section we recall from SGA 4 some basic facts of Grothendieck's theory of cohomology in a topos. In Section 2, we recall the notion of cohomology of algebraic stacks from Laumon and Moret-Bailly [L-MB], and remark, after Behrend [Be1] and Olsson [O], that the smooth-étale site is not functorial. In section 3, we narrate how Laszlo and Olsson ([L-O], [O]) still constructed a pullback in cohomology for algebraic stacks. In Section 4, we collect some well known basic facts about the characteristic classes in étale cohomology for GL_n and $O(n)$.

1.1 Cohomology in a topos

1.1.1 Sites and topoi

We recall the notion of a topos from SGA 4 II.

Definition 1. Let $\mathscr{C}$ be a category. A *Grothendieck topology* on $\mathscr{C}$ attaches to each object X of $\mathscr{C}$ a collection $\mathrm{Cov}(X)$ of families of morphisms to X in such a way that the following hold.
1. The singleton family $(\mathrm{id}_X : X \to X)$ belongs to $\mathrm{Cov}(X)$.
2. If $(u_i : X_i \to X)_i$ lies in $\mathrm{Cov}(X)$, and if for each i, $(y_{i\alpha} : Y_{i\alpha} \to X_i)_\alpha$ lies in $\mathrm{Cov}(X_i)$, then the family $(u_i y_{i\alpha} : Y_{i\alpha} \to X)_{i\alpha}$ lies in $\mathrm{Cov}(X)$.
3. If $f : Y \to X$ is a morphism in $\mathscr{C}$, and if $(X_i \to X)_i \in \mathrm{Cov}(X)$, then each

$X_i \times_X Y$ is representable in $\mathscr{C}$, and the family $(X_i \times_X Y \to Y)_i$ also lies in $\mathrm{Cov}(X)$.

The families in $\mathrm{Cov}(X)$ are called the *coverings* of X.

A *site* is a category together with a Grothendieck topology on it.

Remark 1.1.1. Grothendieck topology, as defined above, (where we follow the common practice of recent decades) is referred to as a *pretopology* in SGA 4 II 1.3 ("prétopologie" in French). Nevertheless, a pretopology generates a unique topology in the sense of SGA 4 II 1.1 and 1.1.6. The above definition of a site therefore coincides with that in SGA 4 II 1.1.5.

Let $\mathscr{C}$ be a site. Let $\mathscr{C}^{\mathrm{op}}$ be the opposite category of $\mathscr{C}$, and let **Sets** be the category of sets.

Definition 2. A *presheaf* (of sets) F on $\mathscr{C}$ is a functor $F : \mathscr{C}^{\mathrm{op}} \to \mathbf{Sets}$. A presheaf (of sets) F is called a *sheaf* (of sets) if for any object X of $\mathscr{C}$, and any covering $(X_i \to X)_i$, the following diagram is an equalizer in **Sets**, where the two parallel arrows are induced by the two projections $X_i \times_X X_j \to X_i$ and $X_i \times_X X_j \to X_j$:

$$F(X) \to \prod_i F(X_i) \rightrightarrows \prod_{ij} F(X_i \times_X X_j).$$

Morphisms of presheaves and sheaves are by definition natural transformations of functors.

Definition 3. A *Grothendieck topos* is a category which is equivalent to the category of sheaves of sets on a site.

Example 1.1.2. Let S be a scheme or an algebraic space. Then the *big étale site* $\acute{E}t(\mathrm{Sp}/S)$ is defined as follows. The objects are morphisms of algebraic spaces $f : X \to S$. A morphism between two objects $u : U \to S$ and $v : V \to S$ is a morphism of algebraic spaces $f : U \to V$ such that $u = v \circ f$. The coverings in this topology are families of morphisms $(X_i \to X)_i$ such that the coproduct morphism is étale and surjective:

$$\coprod_i X_i \to X.$$

Now, consider the full subcategory $\mathrm{\acute{E}t}(S)$ of Sp/S whose objects are étale morphisms $f : X \to S$. Note that if $f : U \to V$ determines a morphism between two objects $u : U \to S$ and $v : V \to S$ of this category, then f itself is an étale morphism. The topology defined above induces a topology on this category in a natural way. Together with this topology, $\mathrm{\acute{E}t}(S)$ is called the *small étale site* of S. The corresponding topos is denoted by $S_{\mathrm{\acute{e}t}}$.

Definition 4. A topology on a category $\mathscr{C}$ is called *subcanonical* if for each object X, the presheaf h_X defined by $h_X(T) = \mathrm{Hom}_{\mathscr{C}}(T, X)$ is a sheaf. If E is the corresponding topos, then by Yoneda lemma, we have a fully faithful functor $h : \mathscr{C} \to E : X \mapsto h_X$.

Example 1.1.3. If X is an algebraic space, then consider the big étale site $\acute{\mathrm{E}}\mathrm{t}(\mathrm{Sp}/X)$. The topology on this is subcanonical, which means that for any object $f : Y \to X$, the presheaf represented by this object is a sheaf in étale topology. The topology on the small étale site $\acute{\mathrm{E}}\mathrm{t}(X)$ is also subcanonical.

1.1.2 Properties of a topos

(a) **Sheafification:** (SGA 4 II 3.4) Let $\mathscr{C}$ be a site and $E = \mathrm{Shv}(\mathscr{C})$ the corresponding topos. Let $\hat{\mathscr{C}}$ be the category of all presheaves on $\mathscr{C}$. Then by the definition of a sheaf, there is a natural fully faithful functor $i : E \hookrightarrow \hat{\mathscr{C}}$. This functor admits a left adjoint, called the sheafification functor, and denoted by $a : \hat{\mathscr{C}} \to E$. If F is a presheaf, then aF is called the *sheafification* of F (or *the sheaf associated* to F).

(b) **Exactness of sheafification:** All small projective and inductive limits are representable in any topos E (SGA 4 II 4.1). In particular, there is a final object e_E and an initial object $\emptyset$. If E is the category of all sheaves on a site $\mathscr{C}$, then to each object of $\mathscr{C}$, a final object of E associates a singleton set, and an initial object of E associates the empty set. The sheafification functor in (a) commutes with projective and inductive limits i.e. it is exact.

(c) **Global sections of a sheaf:** In any topos E, the *global sections* of a sheaf $M \in \mathrm{Ob}(E)$ are elements of the set $\Gamma_E(M) = \mathrm{Hom}_E(e_E, M)$, where e_E is the final object in E. In terms of sections over objects of the underlying site, the set of global sections has the following characterisation: if E is the category of sheaves of sets on a site $\mathscr{C}$, then

$$\Gamma_E(M) = \lim_{\mathscr{C}} M(U).$$

Example 1.1.4. Let X be a scheme or an algebraic space, and consider the small étale site $\acute{\mathrm{E}}\mathrm{t}(X)$. Then h_X is the final object in $X_{\acute{\mathrm{e}}\mathrm{t}}$. If F is a sheaf on this site, then the set of global sections of F is naturally identified with $F(X)$ by Yoneda lemma.

(d) **The constant sheaves:** Let $*$ denote the topos of all sets, which is the category of all sheaves on a single point (the site consisting a single point is uniquely determined up to a unique equivalence). If E is any Grothendieck topos, then there is a unique morphism (see below) of topoi $(u^*, u_*) : E \to *$, where $u_*(F) = \mathrm{Hom}_E(e_E, F)$ (SGA 4 IV 4.3). If X is a set i.e. an object of the topos $*$, then u^*X corresponds to the *constant sheaf* X.

1.1.3 Morphisms of topoi.

Let $\mathscr{C}$ and $\mathscr{C}'$ be two sites, and let $f^{-1} : \mathscr{C} \to \mathscr{C}'$ be a functor. This defines a functor on presheaves

$$f_* : \hat{\mathscr{C}'} \to \hat{\mathscr{C}} : F \to F \circ f.$$

The functor f^{-1} is called a *continuous morphism* if f_* takes sheaves to sheaves, i.e. whenever F is a sheaf on $\mathscr{C}'$, f_*F should be a sheaf on $\mathscr{C}$.

For example, if f^{-1} takes coverings to coverings, then f^{-1} is necessarily continuous, but not conversely: the identity functor id : $\mathrm{Sm}(\mathrm{Sp}/S) \to \text{Ét}(\mathrm{Sp}/S)$ (where $\mathrm{Sm}(\mathrm{Sp}/S)$ is the site with underlying category Sp/S but with coverings jointly surjective, smooth morphism) is continuous, but it does not take coverings to coverings.

If Top$\mathscr{S}$ is a site with underlying category $\mathscr{S}$, where Top designates a topology, we write $\mathscr{S}_{\mathrm{top}}$ for the associated Grothendieck topos i.e. the category of sheaves of sets on the site, following the notation of Laumon and Moret-Bailly [L-MB]. For any continuous morphism of sites $f^{-1} : \mathrm{Top}(\mathscr{S}) \to \mathrm{Top}(\mathscr{S}')$, the functor f_* is left exact i.e. it commutes with all small projective limits. Therefore, it admits a left adjoint $f^* : \mathscr{S}_{\mathrm{top}} \to \mathscr{S}'_{\mathrm{top}}$. We write $f = (f^*, f_*) : \mathscr{S}'_{\mathrm{top}} \to \mathscr{S}_{\mathrm{top}}$ the adjoint pair of functors.

Definition 5. (SGA 4 IV 3.1) Let E and E' be two Grothendieck topoi. A *geometric morphism of topoi* from E to E' is a triple $u = (u^*, u_*, \phi)$, where u^* and u_* are two functors

$$u^* : E \to E' \ , \ u_* : E' \to E$$

and ϕ is an adjunction isomorphism of bifunctors in $X' \in \mathrm{Ob}(E')$, $Y \in \mathrm{Ob}(E)$:

$$\phi : \mathrm{Hom}_E(u^*(X'), Y) \xrightarrow{\sim} \mathrm{Hom}_{E'}(X', u_*(Y)),$$

such that the functor u^* is left exact, i.e. it commutes with finite projective limits. The functor u_* is called the direct image, and the functor u^* is called the inverse image, for the morphism of topoi u. If the adjunction ϕ is clear from the context, then a geometric morphism is denoted simply by (u^*, u_*).

Example 1.1.5. Let $f : X \to Y$ be a morphism of schemes (or algebraic spaces). Then there is a continuous morphism of sites $f^{-1} : \text{Ét}(Y) \to \text{Ét}(X)$, which sends an étale morphism $U \to Y$ to its base change $X \times_Y U \to X$. If F is a sheaf on $\text{Ét}(X)$, then f_*F has the following description. If $U \to X$ is any étale morphism, then $f_*(F)(U) := F(U \times_Y X)$. If G is a sheaf on $\text{Ét}(Y)$, then f^*G is the sheafification of the presheaf on $\text{Ét}(X)$ that attaches to each étale morphism $V \to X$ the inductive limit $\operatorname{colim}_{\mathscr{D}} G(U)$, where $\mathscr{D}$ is the category of all commutative diagrams as follows, where $U \to Y$ is an étale morphism:

$$\begin{array}{ccc} V & \to & U \\ \downarrow & & \downarrow \\ X & \xrightarrow{f} & Y \end{array}$$

The functor f^* is a left exact functor. To see this, it will be sufficient to prove that the category $\mathcal{D}$ is filtering. Since $\mathcal{D}$ has finite products, it is enough to show that it has equalizers.

Let $g, h : U \rightrightarrows W$ be two parallel arrows in $\mathcal{D}$. First, the equalizer of g and h in the category of schemes (or algebraic spaces) over Y is defined by the following Cartesian diagram:

$$\begin{array}{ccc} Z & \to & W \\ \downarrow & & \downarrow^{\Delta_{W/Y}} \\ U & \xrightarrow{(g,h)} & W \times_Y W \end{array}$$

The map $Z \to U$ is an open immersion, because $\Delta_{W/Y}$ is an open immersion, as $W \to Y$ is étale. Therefore, the composite $Z \to U \to Y$ is also étale. By the definition of fibred product, there is a morphism $V \to Z$ over Y. This proves that Z is the equalizer of f and g in $\mathcal{D}$.

Note that this produces a morphism of topoi $(f^*, f_*) : X_{\text{ét}} \to Y_{\text{ét}}$.

Example 1.1.6. Let X be an algebraic space. The natural inclusion (which is a fully faithful functor) $\varepsilon^{-1} : \text{Ét}(X) \to \text{Ét}(\text{Sp}/X)$ is a continuous morphism of sites. At the level of sheaves, the functor ε_* takes a sheaf F on $\text{Ét}(\text{Sp}/X)$ to its restriction to $\text{Ét}(X)$. The left adjoint ε^* has the following description. Let $G \in X_{\text{ét}}$ and $f : Y \to X$ be a morphism of algebraic spaces. Then $\varepsilon^*(G)$ is the sheaf on $\text{Ét}(\text{Sp}/X)$ whose value on $f : Y \to X$ is $(f^*G)(Y)$, where $f^* : X_{\text{ét}} \to Y_{\text{ét}}$ is the pullback. Note that this functor ε^* is exact, because f^* is exact. Therefore $(\varepsilon^*, \varepsilon_*) : (\text{Sp}/X)_{\text{ét}} \to X_{\text{ét}}$ is the corresponding morphism of topoi.

To a morphism of algebraic spaces $f : X \to Y$, however, there may not necessarily correspond a morphism of topoi $(\text{Sp}/X)_{\text{ét}} \to (\text{Sp}/Y)_{\text{ét}}$, because (the big) f^* may not be exact (see 1.2.3 for the counterexample due to Behrend).

1.1.4 Cohomology

Ringed topoi and the derived category

We recall the basics of ringed topoi from SGA 4 IV 12. A *ringed topos* is a pair (E, A), where E is a Grothendieck topos and A is a ring object in E. In this thesis, we will assume that the ring object A is *commutative*. A *morphism of ringed topoi* between two ringed topoi (E, A) and (E', A') is a pair $u = (m, m^{\#})$, where $m = (m^*, m_*) : E \to E'$ is a morphism of topoi, and $m^{\#} : m^*A' \to A$ is a morphism of ring objects in E. By adjunction, it gives a morphism of ring objects $A' \to m_*A$ in E'.

For example, any Grothendieck topos E, with the constant sheaf $\mathbb{Z}_E$, becomes a ringed topos.

Let $\mathrm{Mod}(A)$ be the abelian category of all A-module objects of E. Tensor products and internal homs in the category $\mathrm{Mod}(A)$ exist and have the expected formulation. The usual adjunction relations also hold between internal hom and tensor product (see SGA 4 IV 12).

If $A \to B$ is a ring homomorphism in E, then for each A-module M, the tensor product $M \otimes_A B$ is naturally a B-module. If M is an A-module and P is a B-module, then there is a canonical isomorphism $\mathrm{Hom}_A(M, P) \xrightarrow{\sim} \mathrm{Hom}_B(M \otimes_A B, P)$.

A morphism between two ringed topoi $u = (m, m^{\#}) : (E, A) \to (E', A')$ determines an adjoint pair of additive morphisms $(u^*, u_*) : \mathrm{Mod}(A) \to \mathrm{Mod}(A')$ in the following manner: $u_*F = m_*F$ for an A-module F, and if G is an A'-module, then $u^*G = m^*G \otimes_{m^*A'} A$. Note that in case $\phi : m^*A' \to A$ is a flat morphism of rings (i.e. the functor $\mathrm{Mod}(m^*A') \to \mathrm{Mod}(A) : M \mapsto M \otimes_{m^*A'} A$ is exact), then u^* is an exact functor.

If M is an A-module, then $\mathrm{Hom}_E(e_E, M) = \mathrm{Hom}_A(A, M)$. This defines a left exact additive functor $\Gamma_A : \mathrm{Mod}(A) \to \Gamma_E(A)$-mod, where $\Gamma_E(A)$-mod is the abelian category of all modules over the ring $\Gamma_E(A)$.

By SGA 4 II 6.9, $\mathrm{Mod}(A)$ has enough injectives. In particular, in the derived category $D^+(E, A)$ of cohomologically bounded below complexes, every object is represented by a bounded below complex of injective objects. Let K be an object of $D^+(E, A)$. The i-th hypercohomology of K is the abelian group

$$\mathbb{H}^i(E, K) := \mathrm{Hom}_{D^+(E,A)}(A, K[i]) = \mathrm{Hom}_{D^+(E,A)}(A[-i], K),$$

where A (resp. $A[-i]$) denotes the complex whose only possibly nonzero member is A at 0th (resp. i-th) place. If M is an A-module (which will also denote the complex whose only possibly non-zero member is M at 0th place), then the i-th cohomology of M is

$$H^i(E,M) := \mathbb{H}^i(E,M) = \mathrm{Hom}_{D^+(E,A)}(A[-i],M) = \mathrm{Hom}_{D^+(E,A)}(A,M[i]).$$

In particular, $H^0(E,M) = \Gamma_A(M)$ and the functors $H^i(E,-) : \mathrm{Mod}(A) \to \Gamma_E(A)$-mod form the derived functors of the global section functor Γ_A.

The ringed topos $(E,\mathbb{F}_2)$ associated to any Grothendieck topos E

If A is a ring, then the constant sheaf A_E (see 1.1.2 (e)) on E has a natural structure of a ring object. Thus, on any Grothendieck topos E, there is a natural ring object $\mathbb{F}_{2,E}$ (corresponding to the finite field $\mathbb{F}_2$), which we will henceforth denote simply by $\mathbb{F}_2$.

Any morphism of topoi $(u^*,u_*) : E \to E'$ can be naturally regarded as a morphism of ringed topoi $(E,\mathbb{F}_2) \to (E',\mathbb{F}_2)$, for there is a natural isomorphism of ring objects $u^*\mathbb{F}_2 \to \mathbb{F}_2$. Then (u^*,u_*) determines an adjoint pair of additive functors between the corresponding abelian categories. In this case u^* is exact, so it defines a Δ-functor $u^* : D^+(E',\mathbb{F}_2) \to D^+(E,\mathbb{F}_2)$, which commutes with the shift operators. Therefore, if K is an object of $D^+(E',\mathbb{F}_2)$, then there is an induced pullback map, which is a graded homomorphism

$$u^* : \mathbb{H}^*(E',K) \to \mathbb{H}^*(E,u^*K).$$

In particular, there is a pullback in cohomology

$$u^* : H^*(E',\mathbb{F}_2) \to H^*(E,\mathbb{F}_2).$$

Example 1.1.7. Let $f : X \to Y$ is a morphism of algebraic spaces, and let $(f^*,f_*) : X_{\text{ét}} \to Y_{\text{ét}}$ be the induced morphism of topoi (see 1.1.5). There is an isomorphism of ring objects $f^*\mathbb{F}_{2,Y} \to \mathbb{F}_{2,X}$, giving rise to the map at the level of cohomology

$$H^*_{\text{ét}}(Y,\mathbb{F}_{2,Y}) \to H^*_{\text{ét}}(X,\mathbb{F}_{2,X}).$$

1.1.5 Simplicial Topoi

D-topoi

We now recall the basics of D-topoi from SGA 4 Vbis.

Let D be a small category. Particular examples can be $D = \Delta$ or $D = \Delta^+$, where Δ is the category whose objects are the ordered sets $[n] = \{0 < 1 < \ldots < n\}$, indexed by natural numbers, and $\mathrm{Hom}_\Delta([m],[n])$ consists of all non-decreasing maps between the ordered sets $[m]$ and $[n]$. The category Δ^+ is the subcategory of Δ with the same set of objects but where $\mathrm{Hom}_{\Delta^+}([m],[n])$ consists of all *strictly increasing maps* between $[m]$ and $[n]$.

Let $E \to D$ be a category over a base category D. For an object i of D, let E_i be the subcategory of E whose objects are those objects of E that lie over i, and whose morphisms are those that lie over the identity id_i. A morphism $\eta : x \to y$ in E over a morphism $m : i \to j$ in D is called *Cartesian* if given any object x' of the fibre E_i over i and any morphism $\eta' : x' \to y$ over m in D, there is a unique morphism $\alpha : x' \to x$ in E_i such that $\eta \circ \alpha = \eta'$. Given y and m, a Cartesian morphism $x \to y$ over m, if it exists, is determined up to a unique isomorphism. Fixing m, if for each $y \in \mathrm{Ob}(E_j)$ there exists a Cartesian morphism $x \to y$ over m, then the choice of a Cartesian morphism $m^*y \to y$ over m for each y gives rise to an *inverse image functor* $m^* : E_j \to E_i$.

The category E is *fibred over* D, or $E \to D$ is a *fibred category*, if

1) given any $m : i \to j$ in D, there is a corresponding inverse image functor $m^* : E_j \to E_i$, and if the following holds:

2) if $\eta : x \to y$ over m is Cartesian, and $g : l \to i$ is in D, $x' \in \mathrm{Ob}(E_l)$ and $\eta' : x' \to y$ lies over $m \circ g$, then there is a unique morphism $\alpha : x' \to x$ over g such that $\eta \circ \alpha = \eta'$. (SGA 1 VI 6.11)

In particular, the condition 2) implies that the composite of two Cartesian morphisms is Cartesian.

Now, E is *co-fibred* over D if E^{op} is fibred over D^{op} (SGA 1 VI 10). The inverse image functor in E^{op} for $m : i \to j$ in D (i.e. $m^{\mathrm{op}} : j \to i$ in D^{op}) is called the direct image, and denoted by $m_* : E_i \to E_j$.

If E is both fibred and co-fibred over D, then for $x \in \mathrm{Ob}(E_i)$, $y \in \mathrm{Ob}(E_j)$, and $m : i \to j$,

$$\mathrm{Hom}_{E_i}(x, m^*y) = \mathrm{Hom}_{E/m}(x, y) = \mathrm{Hom}_{E_j}(m_*x, y). \tag{1.1}$$

This gives an adjunction $m_* \dashv m^*$ i.e. m_* is left adjoint to m^*.

Definition 6. Let E be a category fibred and cofibred over D. Then $E \to D$ is called a *D-topos* if

1) each E_i is a Grothendieck topos, and

2) for each $m \in \mathrm{Hom}_D(i,j)$, the adjoint pair (f^*, f_*), where $f^* = m_*$ and $f_* = m^*$ gives rise to a morphism of Grothendieck topoi $f = (f^*, f_*) : E_j \to E_i$, i.e. $f^* = m_*$ is exact.

In particular, if $D = \Delta$ or $D = \Delta^+$, then a D-topos is called a *simplicial topos* or a *strictly simplicial topos*, respectively.

Sheaves on a D-topos: the associated Grothendieck topos

Proposition 1.1.8. (SGA 4 Vbis 1.2.11) *For a D-topos E, let $\underline{\Gamma}(E)$ denote the category $\mathscr{H}om_D(D, E)$ of all global sections of $E \to D$. Then the category $\underline{\Gamma}(E)$ is a Grothendieck topos.*

Notation. $\underline{\Gamma}(E)$ is also called the Grothendieck topos of *sheaves* on the D-topos E.

Remark 1.1.9. An object M of $\underline{\Gamma}(E)$ is the same as an indexed family $(M_i, M(u))_{i,u}$, where, for each object i of D there is an object $M_i \in \mathrm{Ob}(E_i)$, and for each $u \in \mathrm{Hom}_D(i,j)$, there is given a morphism $M(u) : M_i \to M_j$ over u, such that the association $u \mapsto M(u)$ is functorial in the following sense. To give such a collection is equivalent to give a collection $(M_i, \theta_u)_{i,u}$ (or equivalently, by adjunction a collection $(M_i, \phi_u)_{i,u}$) where for each $u \in \mathrm{Hom}_D(i,j)$, $\theta_u : M_i \to u^* M_j$ in E_i (resp., $\phi_u : u_* M_i \to M_j$ in E_j) corresponds to $M(u)$ under equation (1.1), such that the association $u \mapsto \theta_u$ (resp., the association $u \mapsto \phi_u$) is functorial i.e. if $v \in \mathrm{Hom}_D(j,k)$, then $u^*\theta_v \circ \theta_u = \theta_{vu}$ (resp., $\phi_v \circ v_*\phi_u = \phi_{vu}$).

Morphisms between D-topoi and between the associated Grothendieck topoi

Definition 7. If E and E' are two D-topoi, a *morphism of D-topoi* $\Phi : E \to E'$ is a pair of D-functors $\Phi^* : E' \to E$ and $\Phi_* : E \to E'$ with an adjunction $\Phi^* \dashv \Phi_*$ i.e. a bi-functorial isomorphism $\mathrm{Hom}(\Phi^* -, -) \xrightarrow{\sim} \mathrm{Hom}(-, \Phi_* -)$, such that for each object i of D, the restricted pair with the restricted adjunction gives a morphism of topoi $E_i \to E'_i$, i.e. each Φ^*_i is left exact.

Remark 1.1.10. If $\Phi = (\Phi^*, \Phi_*) : E \to E'$ is a morphism of D-topoi, then $\Phi_* : E \to E'$ is a *Cartesian* functor, i.e. it sends inverse images to inverse images.

On the other hand, suppose E and E' are two D-topoi, and $\Phi : E \to E'$ is a Cartesian functor over D, such that for each object i of D, the functor $\Phi_i : E_i \to E'_i$ admits a left adjoint Ψ_i which is exact. Then putting together the left adjoints Ψ_i, one gets D-functor $\Psi : E' \to E$ such that (Ψ, Φ) is a morphism of D-topoi (SGA 4 Vbis 1.2.16).

Proposition 1.1.11. (SGA 4 Vbis 1.2.15) *If $(\Phi^*, \Phi_*) : E \to E'$ is a morphism of D-topoi, then the global section construction $\underline{\Gamma}$ gives rise to a morphism $\underline{\Gamma}(E) \to \underline{\Gamma}(E')$ of Grothendieck topoi.*

Example 1.1.12. Let S be a base scheme. Let D be a small category and $X : D^{\mathrm{op}} \to \mathrm{Sp}/S$ be a functor to the category of algebraic spaces over S. This is denoted by $X_\bullet^+$, where X_q for each object q of D stands for the algebraic space $X(q)$. If $D = \Delta$ (resp. $D = \Delta^+$), then $X_\bullet^+$ is called a *simplicial* (resp. *strictly simplicial*) *algebraic space.*

There is a category $\tilde{X}_{\bullet,\text{ét}}$, whose objects are defined as

$$\mathrm{Ob}(\tilde{X}_{\bullet,\text{ét}}) = \coprod_{q \in \mathrm{ob}(D)} \mathrm{Ob}(X^+_{q,\text{ét}}),$$

and whose morphisms between two objects F and G are defined as follows. If both $F, G \in \mathrm{Ob}(X^+_{q,\text{ét}})$ for some object q of D, then a morphism $F \to G$ in this category is just an element of $\mathrm{Hom}_{X^+_{q,\text{ét}}}(F, G)$. If F is an object of $X^+_{q,\text{ét}}$ and G is an object of $X_{s,\text{ét}}$, then a morphism $F \to G$ in this category would be given by a pair (u, θ_u), where $u : q \to s$ is a morphism in D, and $\theta_u : X(u)^*F \to G$ is a morphism of étale sheaves on X_s. Equivalently, this can be given by a morphism $\phi_u : F \to X(u)_*G$ of étale sheaves on X_q. Set $u^* = X(u)_*$ and $u_* = X(u)^*$. The projection functor $\tilde{X}_{\bullet,\text{ét}} \to D$ makes $\tilde{X}_{\bullet,\text{ét}}$ a fibred and cofibred category over D, with fibre $X^+_{q,\text{ét}}$ over q. Since $(X(u)^*, X(u)_*) : X_{s,\text{ét}} \to X^+_{q,\text{ét}}$ is a morphism of topoi, we see that $\tilde{X}_{\bullet,\text{ét}}$ is a D-topos. The associated Grothendieck topos $\underline{\Gamma}(\tilde{X}_{\bullet,\text{ét}})$ is usually written as $X^+_{\bullet,\text{ét}}$. Note that an object of this topos is a collection of pairs $(M_q, \theta_u)_{q,u}$ (equivalently, a collection of pairs $(M_q, \phi_u)_{q,u}$), where for each object q of D, $M_q \in \mathrm{Ob}(X^+_{q,\text{ét}})$, and there is a functorial association $(u : q \to s) \mapsto (\theta_u : X(u)^*M_q \to M_s)$ of morphisms in $X_{s,\text{ét}}$ (equivalently, a functorial association $(u : q \to s) \mapsto (\phi_u : M_q \to X(u)_*M_s)$ of morphisms in $X^+_{q,\text{ét}}$).

Let $X_\bullet$ and $Y_\bullet$ be two functors $D^{\mathrm{op}} \to \mathrm{Sp}/S$, and $f_\bullet : X_\bullet \to Y_\bullet$ be a natural transformation of functors. This naturally induces a morphism of D-topoi $(f_\bullet^*, f_{\bullet *}) : \tilde{X}_{\bullet,\text{ét}} \to \tilde{Y}_{\bullet,\text{ét}}$ in the following manner. The functor $f_{\bullet *}$ is given fibre-by-fibre. This is Cartesian i.e. it preserves inverse images (note, the inverse image under a morphism u in D is given by push forward of sheaves by the morphisms $X(u) : X_s \to X_q$). At each fibre, it admits an exact left adjoint f_q^*. But by Remark 1.1.10, to give a morphism of D-topoi it is sufficient to give a Cartesian functor of D-categories, with fibre-by-fibre adjoints. This morphism of D-topoi gives rise to a morphism of Grothendieck topoi:

$$(f_\bullet^*, f_{\bullet *}) : X^+_{\bullet,\text{ét}} \to Y^+_{\bullet,\text{ét}}.$$

1.2 Cohomology of algebraic stacks

1.2.1 Algebraic stacks

The following introductory material is taken from Chapters 2, 3 and 4 of Laumon and Moret-Bailly [L-MB].

Let S be a fixed base scheme, which is quasi-separated. In the rest of the chapter, unless mentioned otherwise, all schemes, algebraic spaces and morphisms will be assumed to be over S. Let Aff/S be the category whose objects are affine schemes with morphisms to S (they need not be affine morphisms to S unless S is separated), and where morphisms between two objects are given by morphisms of affine schemes which are compatible with the structure morphisms. Let Aff/S be given the étale topology.

An *S-groupoid* is a category $\mathfrak{X}$ with a functor $a : \mathfrak{X} \to \mathrm{Aff}/S$ (called the *structure morphism*) satisfying the following axioms.

(i) For each morphism $V \xrightarrow{\phi} U$ in Aff/S and object x of $\mathfrak{X}$ such that $a(x) = U$, there is at least one morphism $y \xrightarrow{f} x$ in $\mathfrak{X}$ with $a(f) = \phi$.

(ii) For each diagram $z \xrightarrow{h} x \xleftarrow{f} y$ in $\mathfrak{X}$ with image $W \xrightarrow{\chi} U \xleftarrow{\phi} V$ in Aff/S and for each morphism $W \xrightarrow{\psi} V$ in Aff/S such that $\chi = \phi \circ \psi$, there exists a *unique* morphism $z \xrightarrow{g} y$ in $\mathfrak{X}$ such that $h = f \circ g$ and $a(g) = \psi$.

Note that this means, in particular, that $\mathfrak{X} \to \mathrm{Aff}/S$ is a fibred category, and that any morphism in $\mathfrak{X}$ is Cartesian.

For each object U of Aff/S, we denote by $\mathfrak{X}_U$ the fibre of $\mathfrak{X}$ over U, whose objects are those objects x of $\mathfrak{X}$ such that $a(x) = U$ and whose morphisms are those morphisms f of $\mathfrak{X}$ such that $a(f) = \mathrm{id}_U$. The category $\mathfrak{X}_U$ is a groupoid, i.e. all morphisms are isomorphisms.

For each morphism $V \xrightarrow{\phi} U$ in Aff/S and each object x of $\mathfrak{X}_U$, the axioms (ii) implies that the morphism $y \xrightarrow{f} x$ is unique up to a unique isomorphism. For each $V \xrightarrow{\phi} U$ and each $x \in \mathrm{Ob}\,\mathfrak{X}_U$, we choose once and for all a morphism $y \xrightarrow{f} x$ with $a(f) = \phi$, and denote it by $\phi^*x \to x$ or by $x_V \to x$. Moreover, for any morphism $x' \xrightarrow{g} x$ in $\mathfrak{X}_U$, we denote by ϕ^*g or g_V the unique morphism h in $\mathfrak{X}_V$ which makes the following square commutative:

$$\begin{array}{ccc} \phi^*x' & \longrightarrow & x' \\ {\scriptstyle h}\downarrow & & \downarrow{\scriptstyle g} \\ \phi^*x & \longrightarrow & x. \end{array}$$

A 1-*morphism* of S-groupoids $f : \mathfrak{X} \to \mathfrak{Y}$ is a functor between the underlying categories which is strictly compatible with the structure morphisms of $\mathfrak{X}$ and $\mathfrak{Y}$.

Let $\mathfrak{X}' \xrightarrow{f'} \mathfrak{Y} \xleftarrow{f''} \mathfrak{X}''$ be two 1-morphisms of S-groupoids. The *fibre product* of this is the groupoid $\mathfrak{X}' \times_{f',\mathfrak{Y},f''} \mathfrak{X}''$ defined as follows. The fibre $(\mathfrak{X}' \times_{f',\mathfrak{Y},f''} \mathfrak{X}'')_U$ consists of triples (x', x'', α), where $x' \in \operatorname{Ob} \mathfrak{X}'_U$, $x'' \in \operatorname{Ob} \mathfrak{X}''_U$ and $\alpha : f'(x') \to f''(x'')$ in $\mathfrak{Y}_U$. A morphism from (x'_1, x''_1, α_1) to (x'_2, x''_2, α_2) in the fibre over U is a pair $(x'_1 \xrightarrow{g'} x'_2, x''_1 \xrightarrow{g''} x''_2)$ with g' in $\mathfrak{X}'_U$ and g'' in $\mathfrak{X}''_U$ such that $\alpha_2 \circ f'(g') = f''(g'') \circ \alpha_1$. Finally, for each $V \xrightarrow{\phi} U$ in Aff/S, we have $\phi^*(x', x'', \alpha) = (\phi^* x', \phi^* x'', \phi^* \alpha)$, and $\phi^*(g', g'') = (\phi^* g', \phi^* g'')$.

An *S-stack* is an S-groupoid $\mathfrak{X}$ which satisfies the following conditions.

(i) For each $U \in \operatorname{Ob}(\mathrm{Aff}/S)$, and $x, y \in \operatorname{Ob}(\mathfrak{X}_U)$, the presheaf

$$\underline{\mathrm{Isom}}(x, y) : \mathrm{Aff}_{/U} \to \mathbf{Sets} : (V \to U) \mapsto \operatorname{Hom}_{\mathfrak{X}_V}(x_V, y_V)$$

is a sheaf on Aff/S.

(ii) Given a covering (i.e. covering in étale topology) $(V_i \xrightarrow{\phi_i} U)_i$ in Aff/S, any descent data (x_i, f_{ij}) with respect to this covering is effective.

Given any sheaf of sets X on Aff/S in the étale topology, there is the *associated* S-stack, whose fibre over U is the category (actually a set) whose objects are elements of $X(U)$, and whose morphisms are only id_x for each $x \in X(U)$. This S-stack will also be denoted by X. A stack $\mathfrak{X}$ is called *representable by an algebraic space* if it is the associated S-stack of an algebraic space X. A representable stack is simply called an algebraic space. Let $\mathfrak{X}$ be an S-stack, $U \in \operatorname{Ob}(\mathrm{Aff}/S)$ and $x \in \operatorname{Ob}(\mathfrak{X}_U)$. Then we get a unique 1-morphism $U \to \mathfrak{X}$ taking the object id_U to x, where U is regarded as the associated S-stack. Conversely, any 1-morphism $U \to \mathfrak{X}$ comes uniquely from an object of $\mathfrak{X}_U$. This way, one identifies an object with its corresponding 1-morphism. A 1-morphism of S-stacks $f : \mathfrak{X} \to \mathfrak{Y}$ is called *representable* if for each object U of Aff/S and $x \in \operatorname{Ob}(\mathfrak{X}_U)$ (regarded as a 1-morphism $x : U \to \mathfrak{X}$), the fibre product $\mathfrak{X} \times_{f,\mathfrak{Y},x} U$ is a representable stack. Note that if f is representable, and if $g : Y \to \mathfrak{Y}$ is a 1-morphism from a representable stack Y, then the fibre product $\mathfrak{X} \times_{f,\mathfrak{Y},g} Y$ is also representable. Given a representable 1-morphism f, we will say that f is smooth (resp. étale, surjective, a closed immersion, an open immersion, separated, quasi-compact) if for each $U \in \operatorname{Ob}(\mathrm{Aff}/S)$ and $x \in \operatorname{Ob}(\mathfrak{X}_U)$, the morphism of algebraic spaces $\mathfrak{X} \times_{f,\mathfrak{Y},x} U \to U$ is smooth (resp. étale, surjective, a closed immersion, an open immersion, separated, quasi-compact).

An S-stack $\mathfrak{X}$ is called an *algebraic stack* (also called an Artin stack) if it satisfies the following properties.

(i) The diagonal $\mathfrak{X} \to \mathfrak{X} \times_S \mathfrak{X}$ is representable, separated and quasi-compact (note that this means, in particular, any 1-morphism $X \to \mathfrak{X}$ from an algebraic space is representable).

(ii) There is a smooth, surjective 1-morphism $P : X \to \mathfrak{X}$, where X is an algebraic space.

If $\mathfrak{X}$ is an algebraic stack over S, then a smooth, surjective morphism from an algebraic space to $\mathfrak{X}$ is called an *atlas*.

Note that an algebraic space (i.e. a representable stack) is an algebraic stack.

A 1-morphism of algebraic stacks $f : \mathfrak{X} \to \mathfrak{Y}$ is called *smooth* if for an atlas $X \to \mathfrak{X}$ (therefore for all atlases), the composite 1-morphism $X \to \mathfrak{X} \to \mathfrak{Y}$, which is representable, is smooth.

The following example is important for later use.

Example 1.2.1. Let U be an object of Aff/S and G a group sheaf on $\mathrm{Aff}_{/U}$ in étale topology. A G-torsor on U is a sheaf P on Aff/U in étale topology with an action of G on P satisfying the following property: there is an étale covering $(U' \to U)$ in Aff/S such that $P \times_U U'$ is $G \times_U U'$-equivariantly isomorphic to $G \times_U U'$ (with its own action by multiplication).

Let X be an algebraic space over S and G a sheaf of groups on Aff/S in étale topology, acting on X (not necessarily freely or transitively). Then there is an S-stack $\mathfrak{X} = [X/G]$, called the quotient stack, which is defined as follows. For an object U of Aff/S, G_U will denote the group scheme $G \times_S U$. The objects of $\mathfrak{X}_U$ are pairs (P, f), where P is a G_U-torsor on U, and $f : P \to X$ is a G-equivariant morphism of sheaves on Aff/S. A morphism between (P, f) and (P', f') is given by a G_U-equivariant morphism $g : P \to P'$ satisfying $f'g = f$. There is a canonical 1-morphism $\pi : X \to \mathfrak{X}$ which takes an object $u : U \to X$ to the pair $(G_U, \tilde{u})$, where $\tilde{u} : G_U \to X$ is the G_U-equivariant morphism determined by $u : U \to X$.

If G is a separated, smooth group scheme of finite presentation over S, then $[X/G]$ is an algebraic stack over S with a canonical atlas $\pi : X \to [X/G]$. If X is a smooth algebraic space, then $[X/G]$ is a smooth algebraic stack i.e. the structure morphism $[X/G] \to S$ is smooth.

If $X = S$, so that G acts trivially, the algebraic stack $[S/G]$ is called the *classifying stack for G*, and is denoted by BG. It is a smooth algebraic stack. Note that an object of BG_U is nothing but a principal G-bundle on U locally trivial in étale topology.

1.2.2 The smooth-étale site of an algebraic stack

Let S be a noetherian scheme. All algebraic stacks, all algebraic spaces or schemes will be over S, and all 1-morphisms of algebraic stacks will be over S. In this section, we will recall some basic definitions and results from Chapter 12 of [L-MB].

Definition 8. Let $\mathfrak{X}$ be an algebraic stack. The *smooth-étale site* of $\mathfrak{X}$, denoted by Lis-ét$(\mathfrak{X})$ (in French, called the *lisse-étale site*, and denoted by Lis-ét$(\mathfrak{X})$), is the site defined in the following way.

The underlying category of Lis-ét$(\mathfrak{X})$ has for objects the smooth 1-morphisms (representable) of algebraic stacks $u : U \to \mathfrak{X}$, where U is an algebraic space. Such an object is denoted by (U, u), and called a *smooth-étale open set* of $\mathfrak{X}$. A morphism from (U, u) to (V, v) is an ordered pair (ϕ, α), consisting of a 1-morphism of algebraic spaces $\phi : U \to V$ and a 2-isomorphism $\alpha : u \Rightarrow v \circ \phi$.

The topology on this category is the one generated by the pretopology where $\mathrm{Cov}(U, u)$ is the set of all families of morphisms $((\phi_i, \alpha_i) : (U_i, u_i) \to (U, u))_{i \in I}$ such that the following 1-morphism of algebraic space is étale and surjective.

$$\coprod_{i \in I} \phi_i : \coprod_{i \in I} U_i \to U$$

The corresponding topos is denoted by $\mathfrak{X}_{\text{lis-ét}}$.

Sheaves on the smooth-étale site

If $(U, u) \in$ Lis-ét$(\mathfrak{X})$, any sheaf $F \in \mathfrak{X}_{\text{lis-ét}}$ defines by restriction a sheaf $F_{U,u}$ on the small étale site of U. If $\phi : (U, u) \to (V, v)$ is an arrow in Lis-ét$(\mathfrak{X})$, then for any sheaf $F \in \mathfrak{X}_{\text{lis-ét}}$, there is a morphism of étale sheaves on V, $\theta_{\phi,\alpha} : F_{V,v} \to \phi_* F_{U,u}$, which by adjunction, induces a morphism of étale sheaves on U, $\theta^\sharp_{\phi,\alpha} : \phi^* F_{V,v} \to F_{U,u}$.

Consider the category $\mathscr{C}$ whose objects are all families $(F_{U,u}, \theta_{\phi,\alpha})_{U,u,\phi,\alpha}$ indexed by open sets (U, u) and morphisms $(\phi, \alpha) : (U, u) \to (V, v)$ in Lis-ét$(\mathfrak{X})$, where for each (U, u), $F_{U,u}$ is a sheaf on Ét(U), and for each (ϕ, α), $\theta_{\phi,\alpha} : F_{V,v} \to \phi_* F_{U,u}$ is a morphism of sheaves on Ét(V), satisfying the following conditions.

(i) For each composite morphism $(U, u) \xrightarrow{(\phi,\alpha)} (V, v) \xrightarrow{\psi,\beta} (W, w)$ in Lis-ét$(\mathfrak{X})$,

$$\psi_* \theta_{\phi,\alpha} \circ \theta_{\psi \circ \beta} = \theta_{\psi \circ \phi, \phi^* \beta \circ \alpha} \,.$$

(ii) For each morphism $(\phi, \alpha) : (U, u) \to (V, v)$ in Lis-ét$(\mathfrak{X})$ with $\phi : U \to V$ étale, the morphism of étale sheaves $\theta^\sharp_{\phi,\alpha} : \phi^* F_{V,v} \to F_{U,u}$ is an isomorphism.

Restricting a smooth-étale sheaf to each small étale site gives a functor $\mathfrak{X}_{\text{lis-ét}} \to \mathscr{C}$. Lemma 12.2.1, [L-MB] says that this functor is an equivalence.

Global sections of a sheaf on the smooth-étale site

Recall that in any topos E, the global sections of a sheaf $M \in \mathrm{Ob}(E)$ form the set $\Gamma_E(M) = \mathrm{Hom}_E(e_E, M)$. If $E = \mathfrak{X}_{\text{lis-ét}}$ for an algebraic stack $\mathfrak{X}$, and if M is a smooth-étale sheaf $M \in \mathfrak{X}_{\text{lis-ét}}$, then the set of global sections $\Gamma_E(M)$ has the following description. If $x : X \to \mathfrak{X}$ is a smooth atlas, and if $p_1, p_2 : X \times_{\mathfrak{X}} X \rightrightarrows X$ denote the two projections, then the following diagram is an equalizer, where $p = x \circ p_1 = x \circ p_2$

$$\Gamma_E(M) \to M(X, x) \overset{p_1^*}{\underset{p_{2*}}{\rightrightarrows}} M(X \times_{\mathfrak{X}} X, p).$$

Pullback under a 1-morphism

Suppose $f : \mathfrak{X} \to \mathfrak{Y}$ is a 1-morphism of algebraic stacks. Suppose $G \in \mathfrak{Y}_{\text{lis-et}}$. Let $(U, u) \in \text{Lis-ét}(\mathfrak{X})$. Consider the category $\mathcal{D}$ whose objects are tuples (V, v, f_V, β), where $(V, v) \in \text{Lis-ét}(\mathfrak{X})$, $f_V : U \to V$ is a morphism of algebraic spaces, and $\beta : f \circ u \Rightarrow v \circ f_v$ is a 2-isomorphism. Represented diagrammatically, such a tuple corresponds to the following 2-commutative diagram.

$$\begin{array}{ccc} U & \overset{f_V}{\to} & V \\ u\downarrow & \nearrow\!\!\!\nearrow_{\beta} & \downarrow v \\ \mathfrak{X} & \underset{f}{\to} & \mathfrak{Y} \end{array}$$

A morphisms between (f_v, V, v, β) and (f_w, W, w, γ) is a pair (g, α) such that $g : V \to W$ is a morphism of algebraic spaces, $\alpha : v \Rightarrow w \circ g$ is a 2-isomorphism, $f_w = g \circ f_v$, and the following compatibility condition holds. Consider the horizontal composite 2-isomorphism

$$\alpha * \text{id}_{f_V} : v \circ f_V \Rightarrow w \circ g \circ f_V = w \circ f_w,$$

and the given 2-isomorphism $\beta : f \circ u \Rightarrow v \circ f_V$. The pair (g, α) is required to satisfy the condition that the vertical composite $(\alpha * \text{id}_{f_V}) \bullet \beta$ of the 2-isomorphisms $\beta : f \circ u \Rightarrow v \circ f_v$ and $\alpha * \text{id}_{f_V} : v \circ f_v \Rightarrow w \circ f_w$ should be the 2-isomorphism $\gamma : f \circ u \Rightarrow w \circ f_w$.

Under the equivalence of 1.2.2, the sheaf f^*G on $\text{Lis-ét}(\mathfrak{X})$ is defined by putting

$$(f^*G)_{U,u} = \operatorname*{colim}_{\mathcal{D}} f_v^* G_{V,v}.$$

This defines the inverse image functor $f^* : \mathfrak{Y}_{\text{lis-ét}} \to \mathfrak{X}_{\text{lis-ét}}$.

Push forward under a 1-morphism

Let $F \in \mathfrak{X}_{\text{lis-ét}}$ and $(V, v) \in \text{Lis-ét}(\mathfrak{Y})$. Consider the category $\mathcal{D}'$ of the 2-commutative diagrams, where $(U, u) \in \text{Lis-ét}(\mathfrak{X})$

$$\begin{array}{ccc} U & \stackrel{f_V}{\to} & V \\ u\downarrow & \nearrow\!\!\!\!\nearrow_\beta & \downarrow v \\ \mathfrak{X} & \underset{f}{\to} & \mathfrak{Y} \end{array}$$

with a similar notion of morphisms as in the previous category $\mathcal{D}$. Then define the sheaf (f_*F) on Lis-ét$(\mathfrak{Y})$ by putting

$$(f_*F)_{V,v} = \lim_{\mathcal{D}'} f_{v*}F_{U,u}.$$

This defines the direct image functor $f_* : \mathfrak{X}_{\text{lis-ét}} \to \mathfrak{Y}_{\text{lis-ét}}$.

It can be seen that the inverse image functor f^* is left adjoint to f_*. Therefore f_* is left exact, and f^* is right exact.

Remark 1.2.2. If $\mathfrak{Z} \xrightarrow{g} \mathfrak{Y} \xrightarrow{f} \mathfrak{X}$ are 1-morphisms of algebraic stacks, then the natural map natural map $g^*f^*S \to (f \circ g)^*S$ is an isomorphism. In other words, the pullback is functorial. This fact can be seen in any of the following two ways.

(1) If $f : \mathfrak{X} \to \mathfrak{Y}$ is representable, then there is a continuous morphism of sites

$$f^{-1} : \text{Lis-ét}(\mathfrak{Y}) \to \text{Lis-ét}(\mathfrak{X}) : (\mathrm{U} \to \mathfrak{Y}) \mapsto (\mathrm{U} \times_{\mathfrak{Y}} \mathfrak{X} \xrightarrow{\mathrm{pr}_2} \mathfrak{X}),$$

and the adjoint pair of functors (f^*, f_*) defined above is induced by this continuous morphism (see 1.1.3). However, even if $f : \mathfrak{X} \to \mathfrak{Y}$ is not representable, to each algebraic stack $\mathfrak{X}$ there corresponds a continuous morphism of sites $\iota_{\mathfrak{X}}{}^{-1}$: Lis$(\mathfrak{X}) \to$ Lis-ét$(\mathfrak{X})$ inducing an equivalence of categories at the level of sheaves $\iota_{\mathfrak{X}*} : \mathfrak{X}_{\text{lis-ét}} \xrightarrow{\sim} \mathfrak{X}_{\text{lis}}$, which is functorial in $\mathfrak{X}$ under representable 1-morphisms. Moreover, for any 1-morphism $f : \mathfrak{X} \to \mathfrak{Y}$, not necessarily representable, there is an induced continuous morphism of sites f^{-1} : Lis$(\mathfrak{Y}) \to$ Lis$(\mathfrak{X})$, inducing the adjoint pair (f^*, f_*) of functors on sheaves defined above, via the identification of $\mathfrak{X}_{\text{lis}}$ with $\mathfrak{X}_{\text{lis-ét}}$. The site Lis$(\mathfrak{X})$ is defined following the older version (1988 preprint) of [L-MB]. The underlying category of Lis$(\mathfrak{X})$ has for objects smooth (not necessarily representable) 1-morphisms $u : \mathfrak{U} \to \mathfrak{X}$ of algebraic stacks. A morphism $(u : \mathfrak{U} \to \mathfrak{X}) \to (v : \mathfrak{V} \to \mathfrak{X})$ is a pair (g, α), where $g : \mathfrak{U} \to \mathfrak{V}$ is a 1-morphism of algebraic stacks, and $\alpha : u \Rightarrow v \circ g$ is a 2-isomorphism (i.e. a natural isomorphism of functors). A covering is a family of morphisms $(g_i, \alpha_i) : (u_i : \mathfrak{U}_i \to \mathfrak{X}) \to (u : \mathfrak{U} \to \mathfrak{X})$ such that the resulting 1-morphism $\coprod_i g_i : \coprod_i \mathfrak{U}_i \to \mathfrak{U}$ is a smooth, surjective 1-morphism of algebraic stacks. The continuous morphism $\iota_{\mathfrak{X}}{}^{-1}$: Lis-ét$(\mathfrak{X}) \to$ Lis$(\mathfrak{X})$ is the inclusion functor. If $f : \mathfrak{X} \to \mathfrak{Y}$ is any 1-morphism, then f determines the continuous morphism of sites

$$f^{-1} : \text{Lis}(\mathfrak{Y}) \to \text{Lis}(\mathfrak{X}) : (\mathfrak{U} \to \mathfrak{Y}) \mapsto (\mathfrak{U} \times_{\mathfrak{Y}} \mathfrak{X} \xrightarrow{\mathrm{pr}_2} \mathfrak{X}).$$

(2) Without defining a new site as above, one can directly see that the pullback is functorial. Let $\mathfrak{Z} \xrightarrow{g} \mathfrak{Y} \xrightarrow{f} \mathfrak{X}$ be 1-morphisms of algebraic stacks, not necessarily representable. Let $S \in \mathfrak{X}_{\text{lis-ét}}$. Then there is a natural map $g^* f^* S \to (f \circ g)^* S$. Let $(W, w) \in \text{Lis-ét}(\mathfrak{Z})$. We show that $(g^* f^* S)_{W,w} \to ((f \circ g)^* S)_{W,w}$ is an isomorphism of sheaves on $\acute{\text{E}}\text{t}(W)$, i.e. it is both a monomorphism and an epimorphism. Both are local on W. So, to prove that it is an epimorphism, it is enough to show that given any $\xi \in ((f \circ g)^* S)_{W,w}(W)$, there is an étale cover $W' \to W$, where the restriction of ξ comes from $(g^* f^* S)_{W,w}$. Indeed, since $((f \circ g)^* S)_{W,w}$ is defined as a colimit above, and since the question of epimorphism is local, we can assume that there is some $(U, u) \in \text{Lis-ét}(\mathfrak{X})$ such that ξ comes from a section of $\xi' \in S_U(U)$ under the pullback by a morphism $W \to U$. Let us take an atlas $V \to \mathfrak{Y} \times_{\mathfrak{X}} U$. Let $\widetilde{W}$ be the pullback of V to W. Being smooth over W, $\widetilde{W}$ admits a section over an étale cover W' of W. Then the restriction of ξ to W' is the pullback of the restriction $\xi'|_V$ under the composite $W' \to \widetilde{W} \to V$. But the restriction of $\xi'|_V$ to W' comes from a section of $(g^* f^* S)_{W,w}$. This proves that the natural map is an epimorphism. Similarly, it can be seen that is also a monomorphism.

1.2.3 Issue of the functoriality of the smooth-étale topoi

As Behrend and Gabber pointed out, unlike in the case of étale topoi of algebraic spaces, the inverse image functor (i.e. the pullback) f^* between the smooth-étale topoi is not necessarily left exact, so the adjoint pair (f^*, f_*) may not define a morphism of topoi. Moreover, the problem is already present for the smooth-étale topoi of schemes.

The following example is due to Behrend ([Be1] 5.3.12):

Example 1.2.3. Let $f : Y \hookrightarrow X$ be a closed immersion of smooth varieties over an algebraically closed field k, and let $s \neq 0$ be a regular function on X vanishing along Y. Then multiplication by s determines the map of smooth-étale sheaves $s : \mathcal{O}_X \to \mathcal{O}_X$. This is a monomorphism, i.e. it has trivial kernel. But its pullback $f^* s : f^* \mathcal{O}_X \to f^* \mathcal{O}_X$ is the zero map $\mathcal{O}_Y \to \mathcal{O}_Y$, so $\ker(f^* s) = \mathcal{O}_Y$. Therefore f does not determine a morphism of topoi Lis-ét(Y) $\to$ Lis-ét(X) unlike in the small étale case.

Indeed, since $\mathcal{O}_X$ for any scheme or stack X is representable in the smooth-étale site by $\mathbb{A}^1_X$, the pullback sheaf $f^* \mathcal{O}_X$ on Lis-ét(Y) is representable by $\mathbb{A}^1_Y$. So, the pullback of $s : \mathcal{O}_X \to \mathcal{O}_X$ is represented by the morphism of schemes $\mathbb{A}^1_Y \to \mathbb{A}^1_Y$ over Y, which is multiplication by $s|_Y$.

Remark 1.2.4. However, if the morphism of algebraic stacks $f : \mathfrak{X} \to \mathfrak{Y}$ is *smooth*, then $(f^* F)_{U,u} = F_{U,uf}$. Therefore, f^* is exact in that case. In particular, we have a morphism of ringed topoi

$$(f^*, f_*) : (\mathfrak{X}_{\text{lis-ét}}, \mathbb{F}_{2,\mathfrak{X}}) \to (\mathfrak{Y}_{\text{lis-ét}}, \mathbb{F}_{2,\mathfrak{Y}}).$$

Remark 1.2.5. Let $\mathfrak{X}$ be an algebraic stack and $(f, \alpha) : (U, u) \to (V, v)$ be a morphism in Lis-ét($\mathfrak{X}$). This may not determine a morphism of topoi $U_{\text{lis-ét}} \to V_{\text{lis-ét}}$ unless f is smooth. To give a precise example, if $U \to \mathfrak{X}$ is an object of Lis-ét($\mathfrak{X}$), then the inclusion at zero $U \hookrightarrow \mathbb{A}^1_U$ is a morphism in Lis-ét($\mathfrak{X}$), which, by the example given by Behrend above, does not determine a morphism of corresponding smooth-étale topoi.

1.3 Pullback in the derived category

Since the pullback f^* under a 1-morphism $f : \mathfrak{X} \to \mathfrak{Y}$ is not necessarily exact, we do not directly get a pullback in derived category as in Section 1.1.4, since the 1-morphism f does not define a morphism of topoi. Due to the efforts of Behrend [Be1], Olsson [O] and Laszlo and Olsson [L-O], there is now a construction of a pullback in a suitable triangulated subcategory of $D^+(\mathfrak{X})$. We now briefly recall the construction of a well defined pullback in the cohomologically Cartesian, bounded below derived category of 2-torsion sheaves, following Olsson [Ol].

1.3.1 The two Grothendieck topoi associated to an object of the smooth-étale site

Let $(U, u) \in \text{Ob(Lis-ét}(\mathfrak{X}))$. Then we have a morphism of ringed topoi (see 1.2.4)

$$(u^*, u_*) : (U_{\text{lis-ét}}, \mathbb{F}_{2,U}) \to (\mathfrak{X}_{\text{lis-ét}}, \mathbb{F}_{2,\mathfrak{X}}).$$

There is another morphism of ringed topoi

$$(\varepsilon_U^*, \varepsilon_{U*}) : (U_{\text{lis-ét}}, \mathbb{F}_{2,U}) \to (U_{\text{ét}}, \mathbb{F}_{2,U}),$$

defined as follows. Note that the inclusion functor Ét(U) $\to$ Lis-ét(U) is a continuous morphism of sites. This induces the direct image functor ε_{U*}, which takes a sheaf F on Lis-ét(U) to its restriction to Ét(U). For any sheaf F' on Ét(U), the inverse image $\varepsilon_U^* F'$ has the following description. For any object $g : V \to U$ of Lis-ét(U) (i.e. a smooth morphism of algebraic spaces), $(\varepsilon_U^* F')_{V,g}$ is the sheaf g^*V on Ét(V), where $g^* : U_{\text{ét}} \to V_{\text{ét}}$ is the usual pullback in étale cohomology.

Both ε_U^* and ε_{U*} are exact, and $\varepsilon^*\mathbb{F}_{2,U} = \mathbb{F}_{2,U}$, $\varepsilon_{U*}\mathbb{F}_{2,U} = \mathbb{F}_{2,U}$. Therefore ε_{U*} applied term-wise defines its own derived functor $\varepsilon_{U*} : D^+(U_{\text{lis-ét}}, \mathbb{F}_{2,U}) \to$

$D^+(U_{\text{ét}}, \mathbb{F}_{2,U})$. Therefore ε_{U*} induces a map in hypercohomology with any complex of 2-torsion sheaves K on Lis-ét(U), $\mathbb{H}^*(U_{\text{ét}}, K) \to \mathbb{H}^*(U_{\text{ét}}, \varepsilon_{U*}K)$ (see Section 1.1.4). The exactness of ε_{U*} implies that the map $\mathbb{H}^*(U_{\text{lis-ét}}, K) \to \mathbb{H}^*(U_{\text{ét}}, \varepsilon_{U*}K)$ is an isomorphism. Indeed, the Grothendieck spectral sequence for the composite of the functors ε_* and the global section $\Gamma(U_{\text{ét}}, -)$ degenerates at E_2 level.

1.3.2 The strictly simplicial topoi for the 0-th coskeleton of an atlas

Let $\mathfrak{X}$ be an algebraic stack, and $X \to \mathfrak{X}$ be a smooth atlas. This gives rise to a strictly simplicial algebraic space $X_\bullet^+ = \operatorname{cosk}_0^+(X/\mathfrak{X})$, where X_q is the q-fold fibred product of X over $\mathfrak{X}$. There are two strictly simplicial topoi associated to the atlas $X \to \mathfrak{X}$ defined as follows.

The first strictly simplicial topos $\tilde{X}^+_{\bullet,\text{ét}}$ consists of small étale topoi $X_{q,\text{et}}$ and the morphisms between these topoi induced by the projection morphisms $X(u) : X_j \to X_i$ for strictly increasing $u : [i] \to [j]$ (see 1.1.12).

The second strictly simplicial topos $\tilde{X}^+_{\bullet,\text{sm-ét}}$ consists of topoi $X_{q,\text{sm-ét}}$, and morphisms of topoi induced by the projection morphisms $X(u) : X_j \to X_i$ for strictly increasing $u : [i] \to [j]$. Note that for each strictly increasing $u : [i] \to [j]$, the projection morphism $X(u)$ is a smooth morphism of algebraic spaces, so it does induce a morphism between the corresponding smooth-étale topoi.

While in the first case there also exists the corresponding *simplicial topos* $\tilde{X}_{\bullet,\text{ét}}$ (i.e. allowing projection morphisms $X(u)$ for the degenerate $u : [i] \to [j]$), it is not always possible to define the simplicial topos in the second case. Indeed, in order to define a simplicial topos, we have to give morphisms of topoi corresponding to the degeneracy maps. The morphisms in the simplicial version corresponding to degeneracy maps are actually closed immersions. By Behrend's example (see 1.2.3) closed immersions do not always give rise to morphism of topoi, because their pullbacks may not be exact.

The Grothendieck topoi of sheaves (see 1.1.5) on these two strictly simplicial topoi are denoted by $X^+_{\bullet,\text{ét}}$ and $X^+_{\bullet,\text{lis-ét}}$ respectively. The constant sheaf of rings corresponding to the finite field $\mathbb{F}_2$ on $X^+_{\bullet,\text{ét}}$ is denoted by $\mathbb{F}_{2,X^+_{\bullet,\text{ét}}}$, and that on $X^+_{\bullet,\text{lis-ét}}$, by $\mathbb{F}_{2,X^+_{\bullet,\text{lis-ét}}}$. The corresponding derived categories of cohomologically bounded below complexes of 2-torsion sheaves are denoted by $D^+(X^+_{\bullet,\text{ét}})$ and $D^+(X^+_{\bullet,\text{lis-ét}})$, respectively.

Description of these topoi in terms of underlying sites

The two Grothendieck topoi $X^+_{\bullet,\text{ét}}$ and $X^+_{\bullet,\text{lis-ét}}$ mentioned above can be seen as the topoi of sheaves on two different sites, which are described below.

There is the site $\text{Ét}(X^+_\bullet)$ defined as follows. The objects are étale morphisms $U \to X_q$ from algebraic spaces U for varying q, and a morphism from $(U \to X_q)$ to $(U' \to X_s)$ is a pair (f, ϕ), where $f : U \to U'$ is a morphism of algebraic spaces, and $\phi : [s] \to [q]$ is a strictly increasing map, such that the following diagram commutes.

$$\begin{array}{ccc} U & \xrightarrow{f} & U' \\ \downarrow & & \downarrow \\ X_q & \underset{X(\phi)}{\to} & X_s \end{array}$$

A covering of $(U \to X_q)$ is a family of morphisms indexed by i as follows, where the family $(u_i : U_i \to U)_i$ is an étale covering of U.

$$(u_i, \text{id}_{[q]}) : (U_i \to X_q) \to (U \to X_q)$$

The topos $X^+_{\bullet,\text{ét}}$ is equivalent to the category of sheaves on this site.

The other site $\text{Lis-ét}(\text{X}^+_\bullet)$ has for objects smooth maps $(U \to X_q)$ of algebraic spaces for varying q. Morphisms and coverings in this site are defined exactly as they are above for the site $\text{Ét}(X^+_\bullet)$ in terms of commutative diagrams and étale covers. Then $X^+_{\bullet,\text{lis-ét}}$ is equivalent to the category of sheaves on $\text{Lis-ét}(\text{X}^+_\bullet)$.

Global sections of sheaves on these sites

The global section functors on $X^+_{\bullet,\text{lis-ét}}$ and $X^+_{\bullet,\text{ét}}$ can be easily obtained by the following recipe. For sheaves $F \in X^+_{\bullet,\text{lis-ét}}$ and $G \in X^+_{\bullet,\text{ét}}$, the following diagram are equalizer diagrams for sheaves.

$$\Gamma(X^+_{\bullet,\text{ét}}, G) \to \Gamma(X_{0,\text{et}}, G_0) \rightrightarrows \Gamma(X_{1,\text{et}}, G_1)$$

$$\Gamma(X^+_{\bullet,\text{lis-ét}}, F) \to \Gamma(X_{0,\text{lis-ét}}, F_0) \rightrightarrows \Gamma(X_{1,\text{lis-ét}}, F_1)$$

1.3.3 Comparison of the three Grothendieck topoi

There is a diagram of morphisms of topoi as follows.

$$\mathfrak{X}_{\text{lis-ét}} \overset{(\pi^*,\pi_*)}{\leftarrow} X^+_{\bullet,\text{lis-ét}} \overset{(\varepsilon^*,\varepsilon_*)}{\to} X^+_{\bullet,\text{ét}}$$

The morphism $(\pi^*, \pi_*) : X^+_{\bullet,\text{lis-ét}} \to \mathfrak{X}_{\text{lis-ét}}$

Let G be a sheaf on Lis-ét$(\mathfrak{X})$. Let $\pi_q : X_q \to \mathfrak{X}$ be the projection. Let $G|_{X_q}$ stand for the small étale sheaf G_{X_q,π_q} on X_q. The inverse image functor i.e. the left adjoint π^* takes G to the sheaf on Lis-ét$(X^+_\bullet)$, whose restriction to each X_q is the sheaf $G|_{X_q}$. The pullback morphisms for projections are given by 1.2.2. Since projection morphisms are smooth, by 1.2.4 this is an exact functor.

Let F be a sheaf in $X^+_{\bullet,\text{lis-ét}}$, giving a sheaf F_0 on X, and a sheaf F_1 on $X \times_{\mathfrak{X}} X$, with two restriction homomorphisms $\theta_i : F_0 \to p_{i*}F_1$ for $i = 1, 2$, where p_i are projections $X_1 \rightrightarrows X_0$. Define π_*F as the following equalizer.

$$\pi_*F = \text{Eq}\left(\pi_{0*}\theta_1,\ \pi_{0*}\theta_2 : \pi_{0*}F_0 \rightrightarrows \pi_{1*}F_1\right)$$

The morphism $(\varepsilon^*, \varepsilon_*) : X^+_{\bullet,\text{lis-ét}} \to X^+_{\bullet,\text{ét}}$

The adjoint pair of functors ε^* and ε_* are defined fibre-by-fibre with the induced restriction homomorphisms. Recall that for each q, the adjoint pair $(\varepsilon^*_{X_q}, \varepsilon_{X_q,*})$ (in the notation of 1.3.1) is a morphism of topoi $X^+_{q,\text{lis-ét}} \to X^+_{q,\text{ét}}$. The direct image functor ε_*, which takes a sheaf to the family of its restrictions to the small étale site, commutes with pushforward of sheaves under the projection maps. In the sense of Remark 1.1.10, this is a Cartesian functor over Δ^+, which admits left adjoints fibre-by-fibre. Therefore we get a morphism of Δ^+ topoi, which by Proposition 1.1.11 gives our desired morphism of associated Grothendieck topoi. Explicitly, if F is a sheaf on Lis-ét$(X^+_\bullet)$ given by a sheaf F_q on Lis-ét(X_q) for each q, then ε_*F is the sheaf on Ét$(X^+_\bullet)$ whose restriction to Ét(X_q) is $\varepsilon_{X_q,*}F_q$. On the other hand, if $G \in X^+_{\bullet,\text{ét}}$ is given by the family $G_q \in X^+_{q,\text{ét}}$, then the pullback ε^*G is given by the family $\varepsilon^*_{X_q}G_q \in X^+_{q,\text{lis-ét}}$.

1.3.4 Cartesian sheaves

Cartesian sheaves on an algebraic stack

Definition 9. If $\mathfrak{X}$ is an algebraic stack and F a sheaf on Lis-ét$(\mathfrak{X})$, then F is called *Cartesian* if for any morphism $(f, \alpha) : (U, u) \to (V, v)$ between two objects of Lis-ét$(\mathfrak{X})$, the induced map of small étale sheaves $f^*F_{V,v} \to F_{U,u}$ is an isomorphism.

The constant sheaf $\mathbb{F}_{2,\mathfrak{X}}$ on Lis-ét$(\mathfrak{X})$ is Cartesian. On the other hand, some commonly occurring sheaves of abelian groups such as $\mathbb{G}_m$ are not Cartesian.

If $f : \mathfrak{Y} \to \mathfrak{X}$ is any 1-morphism of algebraic stacks and F is a Cartesian sheaf on $\mathfrak{X}$, then the pullback sheaf f^*F is also Cartesian.

The abelian category of all smooth-étale sheaves of $\mathbb{F}_{2,\mathfrak{X}}$-modules will be denoted by $\mathrm{Mod}(\mathbb{F}_{2,\mathfrak{X}})$. Its objects are all sheaves of abelian groups which are 2-torsion. The full additive subcategory of $\mathrm{Mod}(\mathbb{F}_{2,\mathfrak{X}})$ consisting of all Cartesian sheaves is closed under the formation of kernel, cokernel and extension. This category is denoted by $\mathrm{Mod}_{\mathrm{cart}}(\mathbb{F}_{2,\mathfrak{X}})$.

A complex whose cohomology sheaves are Cartesian is called *cohomologically Cartesian*. If in an exact triangle $K^\bullet \to L^\bullet \to N^\bullet \to$ in $D^+(\mathfrak{X})$, any two of $K^\bullet, L^\bullet$ and $N^\bullet$ are cohomologically Cartesian, then so is the third by five lemma. If $D^+(\mathfrak{X})$ denotes the bounded below derived category of $\mathrm{Mod}(\mathbb{F}_{2,\mathfrak{X}})$, then what we just observed ensures that there is a triangulated full subcategory $D^+_{\mathrm{cart}}(\mathfrak{X})$ of the derived category $D^+(\mathfrak{X})$ consisting of all cohomologically Cartesian complexes.

Cartesian sheaves in $X^+_{\bullet,\text{lis-ét}}$ and $X^+_{\bullet,\text{ét}}$

A sheaf F on $X^+_{\bullet,\text{lis-ét}}$ is given by sheaves $F_i \in X^+_{q,\text{lis-ét}}$ and for each $m : [i] \to [j]$ strictly monotonic, a map of sheaves $\theta_m : X(m)^*F_j \to F_i$. The sheaf F is called *Cartesian* if all the sheaves F_i are Cartesian and the map θ_m is an isomorphism for each strictly monotonic $m : [i] \to [j]$.

Similarly, a sheaf G on $X^+_{\bullet,\text{ét}}$, given by sheaves $G_i \in X^+_{q,\text{ét}}$ and maps $\phi_m : X(m)^*G_j \to G_i$ strictly increasing maps $m : [i] \to [j]$, will be called *Cartesian* if the maps ϕ_m are isomorphism for all strictly increasing m.

The additive subcategory $\mathrm{Mod}_{\mathrm{cart}}(\mathbb{F}_{2,X^+_{\bullet,\text{lis-ét}}})$ (respectively $\mathrm{Mod}_{\mathrm{cart}}(\mathbb{F}_{2,X^+_{\bullet,\text{ét}}})$ consisting of Cartesian abelian 2-torsion sheaves on $X^+_{\bullet,\text{lis-ét}}$ (respectively, on $X^+_{\bullet,\text{ét}}$) is closed under the formation of kernels, cokernels and extensions. The derived categories of bounded below complexes 2-torsion sheaves which are cohomologically Cartesian on $X^+_{\bullet,\text{lis-ét}}$ and $X^+_{\bullet,\text{ét}}$ will be denoted by $D^+_{\mathrm{cart}}(X^+_{\bullet,\text{lis-ét}})$ and $D^+_{\mathrm{cart}}(X^+_{\bullet,\text{ét}})$, respectively.

Comparison of $D^+_{\mathrm{cart}}(\mathfrak{X})$, $D^+_{\mathrm{cart}}(X^+_{\bullet,\text{lis-ét}})$ and $D^+_{\mathrm{cart}}(X^+_{\bullet,\text{ét}})$

We begin by describing the category $Des(X/\mathfrak{X})$ following Olsson [Ol]. Let $p_1, p_2 : X_1 \rightrightarrows X_0$ be the two projections $X \times_{\mathfrak{X}} X \rightrightarrows X$. Then the objects of $Des(X/\mathfrak{X})$ are pairs (G, ι), where G is a 2-torsion abelian sheaf on $\acute{\mathrm{E}}\mathrm{t}(X)$ and $\iota : p_1^*G \xrightarrow{\sim} p_2^*G$ is an isomorphism of sheaves on $\acute{\mathrm{E}}\mathrm{t}(X_1)$, which satisfies the cocycle condition on X_2. Precisely, if $pr_i : X_3 \to X_1$ be the three projections, and if $pr_{23}, pr_{13}, pr_{12} : X_3 \to X_2$ be the three projections, then the two isomorphisms

$$pr_{13}^*(\iota),\ pr_{23}^* \circ pr_{12}^*(\iota) : pr_1^*G \to pr_3^*G$$

on X_2 should be same. The morphisms in $Des(X/\mathfrak{X})$ between (G, ι) and (G', ι') are given by morphisms of sheaves $f : G \to G'$ such that $\iota \circ p_1^*f = p_2^* \circ \iota'$.

Now, any Cartesian sheaf M on $\mathfrak{X}$ defines an object of $Des(X/\mathfrak{X})$ in the following manner. Let $G = M_{X_0}$, the restriction of M to $\acute{\mathrm{E}}\mathrm{t}(X_0)$. Note that, by definition of a Cartesian sheaf, the restriction homomorphisms $\alpha_1 : p_1^* M_{X_0} \to M_{X_1}$ and $\alpha_2 : p_2^* M_{X_0} \to M_{X_1}$ are isomorphisms. Then setting $\iota = \alpha_2^{-1} \circ \alpha_1$, we get an object (G, ι) of $Des(X/\mathfrak{X})$. This defines a functor

$$\mathrm{Mod}_{\mathrm{cart}}(\mathbb{F}_{2,\mathfrak{X}}) \xrightarrow{\sim} Des(X/\mathfrak{X})$$

which is an equivalence of categories (Lemma 4.5, [Ol]).

Again, a Cartesian sheaf F on $X^+_{\bullet,\acute{e}t}$ defines an object of $Des(X/\mathfrak{X})$ in a very similar manner, giving an equivalence of categories (section 4, [Ol])

$$\mathrm{Mod}_{\mathrm{cart}}(\mathbb{F}_{2,X^+_{\bullet,\acute{e}t}}) \xrightarrow{\sim} Des(X/\mathfrak{X}).$$

Finally, the morphism of topoi $\varepsilon^*, \varepsilon_*$ defined in 1.3.3 induces an adjoint pair of additive functors $(\varepsilon^*, \varepsilon_*) : \mathrm{Mod}_{\mathrm{cart}}(\mathbb{F}_{2,X^+_{\bullet,\acute{e}t}}) \to \mathrm{Mod}_{\mathrm{cart}}(\mathbb{F}_{2,X^+_{\bullet,\text{lis-}\acute{e}t}})$, which are quasi-inverses to each other by the definition of Cartesian sheaves. These give the following equivalences of categories.

$$\mathrm{Mod}_{\mathrm{cart}}(\mathfrak{X}) \xrightarrow{\sim} Des(X/\mathfrak{X}) \xleftarrow{\sim} \mathrm{Mod}_{\mathrm{cart}}(X^+_{\bullet,\acute{e}t}) \underset{\varepsilon^*}{\overset{\varepsilon_*}{\leftrightarrows}} \mathrm{Mod}_{\mathrm{cart}}(X^+_{\bullet,\text{lis-}\acute{e}t})$$

The composition across is the same as $\pi^* : \mathrm{Mod}_{\mathrm{cart}}(\mathfrak{X}_{\text{lis-}\acute{e}t}) \to \mathrm{Mod}_{\mathrm{cart}}(X^+_{\bullet,\text{lis-}\acute{e}t})$.

The following Δ-functors are equivalence of triangulated categories

$$\pi^* : D^+_{\mathrm{cart}}(\mathfrak{X}) \to D^+_{\mathrm{cart}}(X^+_{\bullet,\text{lis-}\acute{e}t}), \text{ and}$$

$$\varepsilon^* : D^+_{\mathrm{cart}}(X^+_{\bullet,\acute{e}t}) \to D^+_{\mathrm{cart}}(X^+_{\bullet,\text{lis-}\acute{e}t})$$

with quasi-inverses given by $R\pi_*$ and ϵ_*, respectively (Theorem 4.7, [Ol]).

1.3.5 Construction of pullback in the derived categories

Given a 1-morphism of algebraic stacks $f : \mathfrak{X} \to \mathfrak{Y}$, if $Y \to \mathfrak{Y}$ is any smooth atlas, then the choice of an atlas $X \to \mathfrak{X} \times_{\mathfrak{Y}} Y$ gives rise to a 2-commutative square of strictly simplicial algebraic stacks, where $X^+_\bullet$ and $Y^+_\bullet$ are strictly simplicial algebraic spaces.

$$\begin{array}{ccc} X^+_\bullet & \longrightarrow & \mathfrak{X} \\ {\scriptstyle f_\bullet}\downarrow & & \downarrow{\scriptstyle f} \\ Y^+_\bullet & \longrightarrow & \mathfrak{Y} \end{array}$$

This gives a diagram, where each arrow is a morphism of topoi.

$$\begin{array}{ccccc} X^+_{\bullet,\text{ét}} & \longleftarrow & X^+_{\bullet,\text{lis-ét}} & \longrightarrow & \mathfrak{X}_{\text{lis-ét}} \\ {\scriptstyle f_\bullet}\downarrow & & & & \\ Y^+_{\bullet,\text{ét}} & \longleftarrow & Y^+_{\bullet,\text{lis-ét}} & \longrightarrow & \mathfrak{Y}_{\text{lis-ét}} \end{array}$$

Since $f_\bullet^*$ is exact (see 1.1.12), it defines a Δ-functor

$$f_\bullet^* : D^+_{\text{cart}}(Y^+_{\bullet,\text{ét}}) \to D^+_{\text{cart}}(X^+_{\bullet,\text{ét}}).$$

The pullback $f^* : D^+_{\text{cart}}(\mathfrak{Y}) \to D^+_{\text{cart}}(\mathfrak{X})$ is defined as the composite

$$D^+_{\text{cart}}(\mathfrak{Y}) \xrightarrow{\varepsilon_*\pi^*} D^+_{\text{cart}}(Y^+_{\bullet,\text{ét}}) \xrightarrow{f_\bullet^*} D^+_{\text{cart}}(X^+_{\bullet,\text{ét}}) \xrightarrow{R\pi_*\varepsilon^*} D^+_{\text{cart}}(\mathfrak{X}).$$

Remark 1.3.1. We wish to show that this f^* is independent of the atlases chosen.

First, let $\pi : X \to \mathfrak{X}$ is a smooth atlas and $p : X' \to X$ is a smooth, surjective morphism of algebraic spaces. Write $\pi' : X' \to \mathfrak{X}$ for the resulting atlas. Consider the strictly simplicial algebraic spaces $X'^+_\bullet$ and $X^+_\bullet$ coming from these two atlases and the morphism $p_\bullet : X'^+_\bullet \to X^+_\bullet$. See that the following diagram of functors commutes:

$$\begin{array}{ccc} D^+_{\text{cart}}(\mathfrak{X}_{\text{lis-ét}}) & \xrightarrow{\varepsilon_*\pi^*} & D^+_{\text{cart}}(X^+_{\bullet,\text{ét}}) \\ & {\scriptstyle \varepsilon'_*\pi'^*}\searrow & \downarrow{\scriptstyle p_\bullet^*} \\ & & D^+_{\text{cart}}(X'^+_{\bullet,\text{ét}}) \end{array}$$

As the functors appearing in the diagram are exact, the above can be checked for a single Cartesian sheaf on $\mathfrak{X}$, for which the commutativity follows by definition. All the three functors are equivalences of triangulated categories.

Now, for two different smooth atlases $X'' \to \mathfrak{X}$ and $X \to \mathfrak{X}$, then we can choose the common atlas $X' = X \times_{\mathfrak{X}} X'' \to X \to \mathfrak{X}$. With the above commutative diagram of equivalences, it follows immediately that f^* is independent of the atlas $X \to \mathfrak{X}$. Similarly, if $Y \to \mathfrak{Y}$ and $Y'' \to \mathfrak{Y}$ are two different atlases, then $Y' = Y \times_{\mathfrak{Y}} Y'' \to Y \to \mathfrak{Y}$ is again an atlas. Starting from an atlas $X \to \mathfrak{X} \times_{\mathfrak{Y}} Y$, we can form the fibre product $X' = X \times_Y Y' \to \mathfrak{X} \times_{\mathfrak{Y}} Y$, which is again an atlas. We have two diagrams $X'^+_\bullet \to X^+_\bullet \to \mathfrak{X}$ and $Y'^+_\bullet \to Y^+_\bullet \to \mathfrak{Y}$, where $X_\bullet$, etc. are the resulting (strictly simplicial) 0th coskeleta, and morphisms $X'^+_\bullet \to Y'^+_\bullet$, $X^+_\bullet \to Y^+_\bullet$ and $\mathfrak{X} \to \mathfrak{Y}$ such that the square involving $X'^+_\bullet, X_\bullet, Y'^+_\bullet, Y^+_\bullet$ is commutative. Applying the above commutative diagram to the atlases $X' \to X \to \mathfrak{X}$ and to $Y' \to Y \to \mathfrak{Y}$, and from the commutative square involving $X'^+_\bullet, X^+_\bullet, Y'^+_\bullet, Y^+_\bullet$, it follows that f^* is independent of the choices of atlases.

Remark 1.3.2. It follows from the above that the pullback is functorial. Precisely, if $g : \mathfrak{Y} \to \mathfrak{Z}$ and $f : \mathfrak{X} \to \mathfrak{Y}$ are two 1-morphisms of algebraic stacks then $f^* \circ g^*$ can be naturally identified with $(gf)^*$ as functors.

Remark 1.3.3. In the special case when the 1-morphism $f : \mathfrak{X} \to \mathfrak{Y}$ is smooth (but not necessarily representable), the pullback of sheaves $f^* : \mathfrak{Y}_{\text{lis-ét}} \to \mathfrak{X}_{\text{lis-ét}}$ is exact, and gives rise to a Δ functor $D^+_{\text{cart}}(\mathfrak{Y}_{\text{lis-ét}}) \to D^+_{\text{cart}}(\mathfrak{X}_{\text{lis-ét}})$. This coincides with the one constructed above.

1.4 Characteristic classes of linear algebraic groups

Characteristic classes for an algebraic group are defined in Chapter 2. We collect here some well known basic results about the characteristic classes for the general linear group GL_n and those of the orthogonal group O_n over fields. In what follows, the base is a separably closed field k of characteristic $\neq 2$. For any algebraic spaces X, the étale cohomology $H^*_{\text{ét}}(X, \mathbb{F}_2)$ is written simply as $H^*(X)$, and for any algebraic stack $\mathfrak{X}$, the smooth-étale cohomology $H^*_{\text{lis-ét}}(\mathfrak{X}, \mathbb{F}_2)$ is written simply as $H^*(\mathfrak{X})$. Note that this is consistent in the sense that for any algebraic space X, $H^*_{\text{ét}}(X, \mathbb{F}_2) = H^*_{\text{lis-ét}}(X, \mathbb{F}_2)$.

1.4.1 GL_n: Chern classes

Let us recall the definition of Chern classes due to Grothendieck. The Chern class $c(L) \in H^2(X)$ of a line bundle L on a scheme X is defined as the image of the isomorphism class $[L] \in \text{Pic}(X) = H^1(X, \mathbb{G}_m)$ in $H^2(X, \mu_2)$ under the connecting morphism of the Kummer sequence $1 \to \mu_2 \to \mathbb{G}_m \to \mathbb{G}_m \to 1$. Let E be a vector bundle of rank n over X, and let $p : \mathbb{P}(E) \to X$ be the projective bundle of lines. If $\xi \in H^2(\mathbb{P}(E))$ denotes the Chern class of $\mathcal{O}_{\mathbb{P}(E)}(1)$, then there is a natural isomorphism of free $H^*(X)$-modules

$$H^*(X)[t]/(t^n) \to H^*(\mathbb{P}(E)) : t^i \mapsto \xi^i,\ i = 1, \dots, n-1.$$

Therefore there are unique cohomology classes, called the Chern classes $c_i(E) \in H^{2i}(X)$, satisfying

$$\xi^n + c_1(E)\xi^{n-1} + \cdots + c_n(E) = 0.$$

If E is a line bundle, then $c_1(E) = c(E)$.

The Chern classes of vector bundles over algebraic stacks are defined in the same way.

The ring $H^*(BGL_n)$ is isomorphic to the free polynomial algebra over $\mathbb{F}_2$ generated by the Chern classes. The proof (which we sketch below for completeness) is similar to that of the topological analogue of this fact (see Borel and Hirzebruch [B-H] Section 9).

If E is a vector bundle (i.e. a locally free sheaf) of rank n over an algebraic stack $\mathfrak{X}$ over k, then the full flag bundle of E is denoted by $Fl(E)$. Note that if G_E is the frame bundle of E, which is a principal GL_n-bundle, then $Fl(E)$ is naturally isomorphic to G_E/B, where $B \subset GL_n$ is the Borel subgroup of the upper triangular matrices. The fibration $Fl(E) \to \mathfrak{X}$ factors through projective bundles, showing that $H^*(\mathfrak{X}) \to H^*(G_E/B) = H^*(Fl(E))$ is injective. Applying this to $\mathfrak{X} = BGL_n$ with E the universal vector bundle, so that the smooth atlas $* = \text{Spec } k \to BGL_n = [*/GL_n]$ can be identified with the frame bundle G_E, we see that $H^*(BGL_n) \to H^*(BB) = H^*([*/B])$ is injective. If T is the diagonal maximal torus in GL_n, so that $B = TU$ for the unipotent group U of the strictly upper triangular matrices, then as U is filtered by affine spaces, it follows that the inclusion $T \to B$ induces isomorphism in cohomology $H^*(BB) \to H^*(BT)$. Consequently, the induced map $H^*(BGL_n) \to H^*(BT)$ is an injection. Now, its image lies in the ring of invariants in $H^*(BT)$ under the action of the Weil group $W = S_n$, where $H^*(BT)$ is naturally identified with $\mathbb{F}_2[t_1, \dots, t_n]$, t_i being the Chern classes of the line bundles defined by the diagonal characters. The invariant ring being freely generated by the symmetric polynomials σ_i in degree i, the above yields a graded isomorphism of rings

$$H^*(BGL_n) \to \mathbb{F}_2[\sigma_1, \dots, \sigma_n] : c_i \mapsto \sigma_i.$$

For another argument, see [Be2]

1.4.2 $O(n)$: Stiefel-Whitney classes

A principal $O(n)$-bundle P on a scheme (or an algebraic stack) X over $\mathbb{Z}[1/2]$, is equivalent to giving a vector bundle E of rank n, together with a nondegenerate symmetric bilinear form $q : E \otimes_{\mathcal{O}_X} E \to \mathcal{O}_X$. Indeed, E can be recovered from P as the associated vector bundle under the natural action of $O(n)$ on k^n, and the quadratic form q is induced by the standard quadratic form $\sum x_i^2$ on k^n. Conversely, given a vector bundle with a nondegenerate quadratic form, we can find orthonormal frames étale locally (or over an atlas, if X is an algebraic stack) by Gram-Schmidt orthonormalization. This cannot be done Zariski locally in general, because the Gram-Schmidt process involves taking square roots of functions. We recall below the definition of Stiefel-Whitney classes of $O(n)$ bundles. See, for example, section 5 of [E-K-V].

Let $p : P := \mathbb{P}(E) \to X$ be the projective bundle of lines. A nondegenerate quadratic form q on E determines an isomorphism $E^\vee \xrightarrow{\sim} E$, and therefore an isomorphism of $p^*E^\vee \xrightarrow{\sim} p^*E$. The resulting composite $\mathcal{O}_P(-1) \to p^*E^\vee \xrightarrow{\sim} p^*E \to \mathcal{O}_P(1)$ determines a section of $\mathcal{O}_P(2)$, which is non trivial since q is symmetric. Its nonzero locus will be denoted by U. On U, the above section $\mathcal{O}_P \to \mathcal{O}_P(2)$ is nowhere vanishing, so it determines a reduction of structure group of $\mathcal{O}_P(2)$ to μ_2. Let $\xi \in H^1(U)$ denote the corresponding cohomology class. Then one has isomorphism of $H^*(X)$-modules

$$H^*(X)[t]/(t^n) \xrightarrow{\sim} H^*(U) : t^i \mapsto \xi^i,\ i = 1, \dots, n-1.$$

The Stiefel-Whitney classes $w_i(E, q) \in H^i(X)$ are uniquely defined by the following equation

$$\xi^n + w_1(E)\xi^{n-1} + \cdots + w_n(E) = 0.$$

It is shown by Jardine [J] that the ring $H^*(BO(n))$ is freely generated over $\mathbb{F}_2$ by the Stiefel-Whitney classes.

1.4.3 $GO(n)$: Holla-Nitsure classes

The general orthogonal group scheme GO_n (also called the group scheme of orthogonal similitudes, and denoted by $GO(n)$) over $\mathbb{Z}[1/2]$ is the closed subgroup scheme of GL_n over $\mathbb{Z}[1/2]$, whose R-valued points for any $\mathbb{Z}[1/2]$-algebra R form the following group.

$$GO_n(R) := \{g \in GL_n(R) \colon \exists \lambda \in R^\times,\ {}^t g g = \lambda I_n\} \subset GL_n(R)$$

The character $\sigma : GO_n \to \mathbb{G}_m$ defined by ${}^t g g = \sigma(g) I_n$ is smooth and surjective, and has the orthogonal group scheme $O_n \subset GO_n$ as the kernel. If S is any notherian scheme over $\mathbb{Z}[1/2]$ then the base change of GO_n to S is denoted by $GO_{n,S}$. However, we will denote it simply by GO_n if the base is understood beyond confusion.

Note that the homomorphism $\mathbb{G}_m \times SO_n \to GO_n : (\alpha, g) \mapsto \alpha g$ is an isomorphism if n is odd, but it has a nontrivial kernel isomorphic to μ_2 if n is even.

The group scheme GO_n is connected if n is odd, but it has two connected components if n is even. For n even, say $n = 2m$, GO_n has the nontrivial character $g \mapsto \det(g)\sigma(g)^{-m}$, which takes the values ± 1. The kernel of this character is the image of the homomorphism $\mathbb{G}_m \times SO_{2m} \to GO_{2m} : (\alpha, g) \mapsto \alpha g$, hence connected.

Since $\sigma : GO_n \to \mathbb{G}_m$ is smooth, and since $\mathbb{G}_m$ is smooth over the base, GO_n is also smooth over the base. The defining representation of SO_n is irreducible.

Therefore the defining representation of GO_n (and also that of its identity component) is irreducible. Therefore (the identity component of) GO_n is reductive in the following sense: a group scheme G over S is reductive if G is affine and smooth over S, and if for each $s \in S$, the geometric fibre $G_{\bar{s}}$ is reductive in the usual sense.

Since $GO_{2n+1} \cong \mathbb{G}_m \times SO_{2n+1}$, we have an isomorphism

$$X^*(GO_{2n+1}) \cong X^*(\mathbb{G}_m) \oplus X^*(SO_{2n+1}) = \mathbb{Z}.$$

In the even case, the characters det and $\det/\sigma^n$ of GO_{2n} define a map

$$\mathbb{Z} \oplus \mathbb{Z}/2\mathbb{Z} \to X^*(GO_{2n}) : (1, \bar{0}) \mapsto \det,\ (0, \bar{1}) \mapsto \det/\sigma^n.$$

It can be seen that this is an isomorphism.

In terms of line bundle valued quadratic forms on vector bundles, GO_n has the following description. A principal GO_n-bundle on a scheme (or stack) X over $\mathbb{Z}[1/2]$ is equivalent to a triple (E, L, b), where E is a vector bundle of rank n, L a line bundle and $b : E \otimes E \to L$ is a nondegenerate symmetric form. Indeed, let E be the associated vector bundle under the defining representation of GO_n, and let L be induced by the character σ, then the symmetric form on the vector bundle is induced by the standard bilinear form $\sum_i x_i y_i$, which is preserved upto scalars by GO_n. Conversely, given a triple (E, L, b), local orthonormal frames can be constructed using Gram-Schmidt orthogonalization étale locally (or over an atlas if X is an algebraic stack).

Let k be a separably closed field of characteristic different from 2. We give below a brief description of the characteristic classes for principal $GO_{n,k}$-bundles in étale cohomology, which are fully explained in Chapters 2 and 3. In what follows, the group schemes SO_n, etc. would stand for the corresponding group schemes $SO_{n,k}$, etc.

It is well known that $H^*(BSO_{2n+1}, \mathbb{F}_2) = \mathbb{F}_2[w_2, \dots, w_{2n+1}]$. The isomorphism of group schemes $\mathbb{G}_m \times SO_{2n+1} \to GO_{2n+1}$ provides an isomorphism of rings $BGO_{2n+1} \xrightarrow{\sim} B\mathbb{G}_m \times BSO_{2n+1}$. Now, $H^*(B\mathbb{G}_m, \mathbb{F}_2) = \mathbb{F}_2[\bar{c}_1]$ where $\bar{c}_1$ is the first Chern class modulo 2. Hence by Künneth formula,

$$H^*(BGO_{2n+1}, \mathbb{F}_2) \cong \mathbb{F}_2[c, w_2, \dots, w_{2n+1}]$$

where c denotes the pullback of the first Chern class $\bar{c}_1$ from $H^2(BG_m, \mathbb{F}_2)$, and the $w_2, \dots, w_{2n+1}$ are the pull-backs of the Stiefel-Whitney classes in $H^*(BSO_{2n+1}, \mathbb{F}_2)$.

In the even case, the ring $H^*(BGO_{2n}, \mathbb{F}_2)$ is generated by the cohomology classes λ, a_{2i-1}, b_{4j}, and d_T over $\mathbb{F}_2$, where $i, j \in \{1, \dots, n\}$ and $T \subset \{1, \dots, n\}$ has cardinality $|T| \geq 2$. These are defined as follows. Let (E_{2n}, L_{2n}, b_{2n}) denote the universal quadratic triple on BGO_{2n}. Then $\lambda = \bar{c}_1(L_{2n})$ and $b_{4j} = \bar{c}_{2j}(E_{2n})$. Next, consider the Gysin boundary map $d : H^*(BO_n, \mathbb{F}_2) \to H^{*-1}(BGO_n, \mathbb{F}_2)$ that we introduced earlier. Then $a_{2i-1} = d(w_{2i}) \in H^{2i-1}(BGO_n, \mathbb{F}_2)$, and for any $T = \{i_1, \ldots, i_r\} \subset \{1, \ldots, n\}$, we have $d_T = d(w_{2i_1} \cdots w_{2i_r}) \in H^{2(i_1+\ldots+i_r)-1}(BGO_n, \mathbb{F}_2)$. These generators were originally introduced by Holla and Nitsure in [H-N-1] and [H-N-2]; their relations are given explicitly in Theorem 3.9 of [H-N-1]. In the algebraic case they satisfy precisely the same relations. In other words, the $\mathbb{F}_2$-algebra $H^*(BGO_{2n}, \mathbb{F}_2)$ has the same description in terms of the generators λ, a_{2i-1}, b_{4j}, d_T and relations as the singular cohomology algebra $H^*_{\mathrm{sing}}(BGO(2n, \mathbb{C}), \mathbb{F}_2)$. This fact is the main result of Chapter 2 of this thesis. The ring homomorphisms $H^*(BGL_{2n}, \mathbb{F}_2) \to H^*(BGO_{2n}, \mathbb{F}_2) \to H^*(BO_{2n}, \mathbb{F}_2)$ induced by the inclusions $O_{2n} \hookrightarrow GO_{2n} \hookrightarrow GL_{2n}$ have the same concrete description in terms of the above generators as given in Section 3 of [H-N-2].

Chapter 2

Characteristic classes for $GO(2n)$

2.1 Introduction

Let k be an algebraically closed field of characteristic $\neq 2$. The **general orthogonal group**, also known as the **group of similitudes**, is the closed subgroup scheme $GO(n) \subset GL_{n,k}$ whose set of R-valued points, for any k-algebra R, is

$$GO(n)(R) = \{A \in GL_n(R) : \exists a \in R^\times, \ {}^tAA = aI_n\}.$$

This is a reductive group scheme, since it is reduced and its defining representation on k^n is irreducible. In the topological category, Y. Holla and N. Nitsure [H-N-1] determined the characteristic classes for the complex Lie group $GO(n)$ with coefficients in $\mathbb{F}_2$, i.e. they explicitly determined the singular cohomology ring $H^*(BGO(n); \mathbb{F}_2)$ in terms of generators and relations. The present note proves a similar result in the algebraic category for the étale cohomology for the algebraic stack $BGO(n)$.

There are two basic components in our extension of the methods of [H-N-1] to the algebraic category. The first of them is a good functorial description of the cohomology of the stack $BGO(n)$. This can be achieved by extending to étale cohomology an idea of Totaro [Tot], which he originally used for treating Chow groups. The second component is the Gysin sequence for the $\mathbb{G}_m$-torsor $BO(n) \to BGO(n)$ in étale cohomology, which is actually a by-product of the first component. Once these points are established, the determination of the ring of characteristic classes proceeds exactly as in [H-N-1].

As was remarked in section 1 of [H-N-1], the group scheme $GO(2n+1)$ is a product $\mathbb{G}_m \times SO(2n+1)$, so $BGO(2n+1) \cong B\mathbb{G}_m \times BSO(2n+1)$, and

$$\begin{aligned} H^*_{et}(BGO(2n+1); \mathbb{F}_2) &\cong H^*_{et}(B\mathbb{G}_m; \mathbb{F}_2) \otimes H^*_{et}(BSO(2n+1); \mathbb{F}_2) \\ &= \mathbb{F}_2[\lambda] \otimes \mathbb{F}_2[w_2, \ldots, w_{2n+1}] = \mathbb{F}_2[\lambda, w_2, \ldots, w_{2n+1}]. \end{aligned}$$

It is perhaps well known that $H^*_{et}(BSO(2n+1);\mathbb{F}_2) \cong \mathbb{F}_2[w_2,\ldots,w_{2n+1}]$. Nonetheless, we sketch a proof of this fact at the end of Section 2.5 for the reader's reference. Thus, the case of interest is the cohomology ring of $BGO(2n)$.

2.2 Notations and preliminaries

Henceforth, all schemes considered in this note will be quasi-projective over an algebraically closed field k of characteristic $\neq 2$, and all morphisms will be over k. For such a scheme X, we will denote $H^*_{et}(X;\mathbb{F}_2)$ simply by $H^*(X)$ in what follows. A morphism $X \to Y$ of schemes will be called n-**acyclic**, if the induced map $H^i(Y) \to H^i(X)$ is a bijection for $i \leq n$. A scheme X will be called n-**acyclic** if the structure morphism $X \to \operatorname{Spec} k$ is n-acyclic. If a morphism is n-acyclic for all n, it will be called an **acyclic** morphism.

All algebraic stacks considered in this note will be over k, and all 1-morphisms will be over k. We will denote the smooth-étale cohomology $H^*_{sm\text{-}et}(\mathfrak{X};\mathbb{F}_2)$ simply by $H^*(\mathfrak{X})$. Note that if $\mathfrak{X}$ is representable by an algebraic space, then $H^*_{et}(\mathfrak{X};\mathbb{F}_2) = H^*_{sm\text{-}et}(\mathfrak{X};\mathbb{F}_2)$. A 1-morphism of algebraic stacks $X \to \mathfrak{X}$, where X is representable by an algebraic space, will be called n-**acyclic** if the map induced in cohomology $H^i(\mathfrak{X}) \to H^i(X)$ is a bijection for $i \leq n$.

For any smooth group scheme G over k, BG will denote the algebraic stack of all principal bundles on the category of quasi-projective k-schemes, locally trivial in étale topology. A **characteristic class** for G in étale cohomology with coefficient $\mathbb{F}_2$ is a natural transformation $|BG| \to H^*_{et}(\;;\mathbb{F}_2)$ of functors from the category of quasi projective schemes over k to the category of sets, where $|BG|$ is defined by taking $|BG|(X)$ to be the set of isomorphism classes in $BG(X)$ for any quasi projective scheme X. All such characteristic classes for G form a ring, whose addition and multiplication come from the addition and cup product structure on H^*.

Definition 10. Let G be a group scheme over k, acting on a k-scheme X. A **bundle quotient** for this action is a morphism $\phi : X \to Y$ of k-schemes, such that X together with the given G-action is a principal G-bundle on Y, locally trivial in the étale topology.

Remark 2.2.1. Clearly, whenever such a bundle quotient exists, it is unique up to a unique isomorphism. The existence of bundle quotient is the same as the representability of the quotient stack $[X/G]$ by a scheme. In what follows, bundle quotients will naturally occur in two different ways:
a) If a reductive group G acts on a finite type, affine scheme X, and $U \subset X$ is an invariant open subscheme such that for each closed point $x \in U$, the orbit

$O(x)$ is closed in X, and the induced action of G on U is free, then U admits a bundle quotient by G. Indeed, since G is reductive, we have a uniform categorical quotient $\phi : X \to X/G$ with X/G affine and ϕ universally submersive (Theorem 1.1, [GIT]). As U is invariant, its image $\phi(U) \subset X/G$ is open. It follows that $\phi : U \to \phi(U)$ is a geometric quotient. Again, since the action on U is free, $\phi : U \to \phi(U)$ is a principal G-bundle (Proposition 0.9 of [GIT]).
b) An affine algebraic group scheme G has a bundle quotient by any of its reduced closed subgroup schemes H, which acts on G by right translations. If G is reduced, hence smooth over k, then G/H, being a reduced homogeneous space, is smooth over k (k being algebraically closed). As H is reduced, hence smooth over k, $G \to G/H$ is étale locally trivial.

Note that in each of the above cases, the quotient space is a quasi-projective scheme over k.

2.3 Characteristic classes for G and the cohomology of the stack BG

In this section, cohomology will mean étale cohomology with coefficient $\mathbb{Z}/n\mathbb{Z}$ or any constant torsion coefficient such that n or the torsion is not divisible by the characteristic of the ground field k.

There is a natural map Θ_G from the cohomology of the classifying stack BG to the ring of characteristic classes, which takes a cohomology class $\nu \in H^*(BG)$ to the characteristic class whose value on a principal G-bundle $P \to X$ is $f^*\nu$, where $f : X \to BG$ is the classifying morphism of P. Our treatment of this map is inspired by the article Totaro [Tot], which introduced a similar map (actually the inverse of Θ_G) for chow groups, and proved that it was a bijection.

Proposition 2.3.1. *When G is a reductive group scheme over k, the map Θ_G is a bijection.*

Before we write down the proof of this theorem, let us recall some facts and make a few observations.

Let $Z \hookrightarrow X$ be a closed immersion of two smooth schemes over k, where X has relative dimension n over k, and all the irreducible components of Z have codimension at least $(s+1)$ in X. Then each point $z \in Z$ has an open neighbourhood $z \in U \subset X$, and there is an étale morphism $\pi : U \to \mathbb{A}^n_k$ such that $Z \cap U = \pi^{-1}(\mathbb{A}^{n-t}_k)$ with $t \geq s+1$. Then the theorem of cohomological purity (Theorem 5.1 of [M]) shows that $\underline{H}^i_Z(X) = 0$ for $i < 2s+2$. From the spectral sequence $H^i(Z, \underline{H}^j_Z(X)) \Rightarrow H^{i+j}_Z(X)$, which shows that $H^i_Z(X) = 0$ for $i < 2s + 2$, and the following long exact sequence

$$\cdots \to H^i_Z(X) \to H^i(X) \to H^i(X - Z) \to H^{i+1}_Z(X) \cdots$$

we conclude that the open immersion $(X - Z) \hookrightarrow X$ is $2s$-acyclic. However, we need a little stronger statement, where Z need not be smooth:

Lemma 2.3.2. *Let $Z \hookrightarrow X$ be a closed immersion of finite type schemes over k with X smooth over k. If all the irreducible components of Z have codimension $\geq s + 1$ in X, then the open immersion $(X - Z) \hookrightarrow X$ is $2s$-acyclic.*

Proof. The underlying reduced subscheme Z_{red} has a finite filtration by reduced, closed subschemes

$$Z_{red} = Z_0 \supset Z_1 \supset \cdots \supset Z_\ell,$$

where for each $i \geq 1$, Z_i is the singular locus of Z_{i-1} and Z_ℓ is non-empty, smooth over k. Then $\mathrm{codim}(Z_i, X) \geq s + i + 1$ for each i. As Z_ℓ is smooth over k, $X - Z_\ell \hookrightarrow X$ is $(2s + 2l)$-acyclic by the paragraph preceding the Lemma. Note that $Z_{i-1} - Z_i$ is a smooth over k, closed subscheme of $X - Z_i$ of codimension $\geq s+i+1$, so that $X - Z_{i-1} \hookrightarrow X - Z_i$ is $(2s+2i)$-acyclic for each $i \geq 1$, by the same argument. Therefore, $X - Z \hookrightarrow X$ is $2s$-acyclic. □

Lemma 2.3.3. *Let $P \to X$ be a principal G-bundle over k with G reduced and X quasi-projective over k. Let f be its classifying morphism $X \to BG$. If P is n-acyclic over k, then the morphism f is also n-acyclic.*

Proof. We have to show that $R^i f_* \mathbb{F}_{2,X} = 0$ where $\mathbb{F}_{2,X}$ is the constant sheaf on X. Let $\pi : \mathrm{Spec}\, k \to BG$ be the classifying morphism of the trivial bundle $G \to \mathrm{Spec}\, k$. As π is smooth and surjective, to show that $R^i f_* \mathbb{F}_{2,X} = 0$, it is enough to show that $\pi^* R^i f_* \mathbb{F}_{2,X} = 0$. Note that we have a Cartesian square

$$\begin{array}{ccc} P & \longrightarrow & X \\ {\scriptstyle f'}\big\downarrow & & \big\downarrow{\scriptstyle f} \\ \mathrm{Spec}\, k & \xrightarrow{\pi} & BG \end{array}$$

By smooth base-change, $\pi^* R^i f_* \mathbb{F}_{2,X} = R^i f'_* \mathbb{F}_{2,P} = H^i(P)$, which is zero by hypothesis. □

A morphism $X' \to X$ will be called an **affine space bundle** if X admits an étale cover $\{X_i \to X\}_{i \in I}$ such that each base change $X' \times_X X_i \to X_i$ is an affine space over X_i.

Lemma 2.3.4. *Let X be a quasi-projective scheme over k, E a principal G-bundle on X, where G is reductive over k, and $s > 0$ an integer. Then there is an affine space bundle $\pi : X' \to X$, an affine space V over k with a linear G-action on it, and a closed subset $S \subset V$ such that G acts freely on $(V - S)$ and the pullback π^*E is isomorphic as a G-bundle over X' to the pullback of the principal G-bundle $(V - S) \to (V - S)/G$ by a morphism $f : X' \to (V - S)/G$. Moreover, the quotient space $(V - S)/G$ is a quasi-projective scheme over k, and the classifying morphism $(V - S)/G \to BG$ for the principal G-bundle $(V - S) \to (V - S)/G$ is $2s$-acyclic.*

Proof. First see that if V is an affine space with a linear G-action, and if S is a closed subset of V such that the orbit of each closed point in $(V - S)$ is closed in V, and G acts freely on $(V - S)$, then by Remark 2.2.1 a), we have a bundle quotient $(V - S) \to (V - S)/G$. Therefore, the first statement follows from Lemma 1.6 of [Tot] (and the proof of it). The quotient space $(V - S)/G$ is quasi-projective by Remark 2.2.1. By Lemma 2.3.3, we only have to show that the inclusion $(V - S) \hookrightarrow V$ is $2s$-acyclic, as V is acyclic over k. The proof of Lemma 1.6 of [Tot] shows that V and S can be so chosen that S has codimension $\geq s + 1$ in V. □

Lemma 2.3.5. *Let G be any reduced algebraic group over k. Then, given any positive integer s, there exists a linear representation V of G, a closed subset S of V such that G acts freely on $(V - S)$, the bundle quotient $(V - S) \to (V - S)/G$ exists, the total space $(V - S)$ is $2s$-acyclic over k, and the quotient space is smooth, quasi-projective over k. In particular, the classifying morphism of this principal G-bundle is $2s$-acyclic.*

Proof. This is precisely Remark 1.4 of [Tot]. To recall the same, let G be regarded as a closed subgroup scheme of $GL_{n,k}$ for some n. For any positive integer s, choose a positive integer p such that $(p - n) \geq s$. Let $M^o_{n,p}$ be the scheme of $n \times p$ matrices of rank n, regarded as an open subscheme of $M_{n,p} = \mathbb{A}^{n \times p}$. Recall that for the left action of $GL_{n,k}$ on $M^o_{n,p}$, the bundle quotient is the Grassmannian $Gr(k^p, n)$ of n-dimensional quotients of k^p, and $M^o_{n,p} \to Gr(k^p, n)$ is the tautological principal GL_n-bundle on the Grassmannian. Now, the total space of the associated GL_n/G-bundle on $Gr(k^p, n)$ is just the bundle quotient of $M^0_{n,p}$ by the action of G, which exists as a smooth, quasi-projective scheme over k because it is a quotient of $GL_{p,k}$ by a smooth, closed subgroup scheme (Remark 2.2.1 b)). Also observe that the codimension of the complement of $M^o_{n,p}$ in $M_{n,p}$ is $\geq (n - p + 1) \geq (s + 1)$. Then by Lemma 2.3.2, we can take $V = M_{n,p}$, and $(V - S) = M^o_{n,p}$. The last sentence follows from Lemma 2.3.3. □

Proof of Proposition 2.3.1. The injectivity is immediate: if $\Theta_G(\beta) = 0$ for a nonzero $\beta \in H^i(BG)$ then by Lemma 2.3.5 we find a principal G-bundle $P \to X$, whose classifying morphism $X \to BG$ is i-acyclic, yielding a contradiction to the definition of Θ_G.

To prove the surjectivity, starting off from a characteristic class ν of homogeneous degree s, say, we choose any pair (V, S_V), where V is a linear representation of G, and S_V is a closed subset of V of codimension $\geq s+1$, such that G acts freely on $(V - S_V)$, so that the classifying morphism $(V - S_V)/G \to BG$ is $2s$-acyclic. The value of ν on this principal G-bundle is the image, under the induced map in cohomology, of a unique element, say $\alpha \in H^s(BG)$. We claim that $\nu = \Theta_G(\alpha)$. The claim will follow immediately from Lemma 2.3.4, once we show that α is independent of the pair (V, S_V). But this in turn is a consequence of Totaro's argument involving "independence of V and S", which we recall from the proof of Theorem 1.1 of [Tot]: let (W, S_W) be another pair such as (V, S_V). Then the bundle quotients $((V - S_V) \times W)/G$ and $(V \times (W - S_W))/G$ exist as vector bundles over $(V - S_V)/G$ and $(W - S_W)/G$, respectively. Both the closed subsets $V \times S_W$ and $S_V \times W$ of $V \oplus W$ have codimensions at least $(s+1)$, and outside each of them, G acts freely, so that in each case the bundle quotient exists. Therefore, there is an invariant open subscheme of $V \oplus W$ which contains both of these open complements and which consists of G-stable points. Let us call this open subscheme $V \oplus W - S_{V \oplus W}$. Observe that $S_{V \oplus W}$, being contained in each of the two closed subsets mentioned above, has codimension at least $(s+1)$ in $V \oplus W$. Since $H^i((V - S)/G)$ for $i \leq 2s$ does not depend on S as long as the codimension of S is $\geq s+1$, we see that all the arrows in the following commutative diagram are isomorphisms for $i \leq 2s$, where the right vertical, the diagonal, and the lower horizontal arrows are induced by the classifying morphisms, and the upper horizontal and the left vertical arrows, by obvious inclusions:

$$\begin{array}{ccc} H^i((V \oplus W - S_{V \oplus W})/G) & \longrightarrow & H^i((V \times (W - S_W))/G) = H^i((W - S_W)/G) \\ \downarrow & \nwarrow & \uparrow \\ H^i((V - S_V)/G) = H^i(((V - S_V) \times W)/G) & \longleftarrow & H^i(BG) \end{array}$$

This completes the proof. □

Remark 2.3.6. The independence of V and S as above can also be applied in a little different setup: suppose $P \to X$ is a principal G-bundle, and there are two G-equivariant pairs (V, S_V) and (W, S_W) with maps from X to the bundle quotients $(V - S_V)/G$ and $(W - S_W)/G$, with properties as required in Lemma 2.3.4. Then the diagonal map $P \to V \oplus W$ factors through an open inclusion $(V \oplus W - S_{V \oplus W}) \hookrightarrow V \oplus W$ as above.

2.4 The Gysin sequence

We begin by observing the following:

Remark 2.4.1. Let $L \to X$ be a line bundle over k. A line bundle being locally trivial in the Zariski topology, the zero section has *pure codimension* in L. Therefore, whenever there is a principal $\mathbb{G}_m$-bundle with the quotient space, hence the total space as well, smooth over k, we have the Gysin sequence for the smooth pair (L, X), where X is regarded as the zero section.

Lemma 2.4.2. *Let $1 \to N \to G \to G' \to 1$ be an exact sequence of reduced, affine algebraic groups over k, and let $E \to X$ be a principal G-bundle. Then the morphism $E \to X$ admits a factorization $E \to E' \to X$, where $E \to E'$ is the bundle quotient of E by N, and $E' \to X$ is the G'-bundle associated to $E \to X$.*

Proof. The associated G'-bundle $E' \to X$ exists by étale descent of affine morphisms. It is seen locally that the morphism $E \to E'$ is the required bundle quotient. □

In the above, let $G' = \mathbb{G}_m$, the multiplicative group scheme. Let $P \to Q$ be a principal G-bundle, where Q is a smooth, quasi-projective scheme over k. Then the bundle quotient of P by the action of N exists as the $\mathbb{G}_m$-bundle $Q' \to Q$ associated to $P \to Q$, by Lemma 2.4.2. We can recognise $Q' \to Q$ as the complement of the zero section of the induced line bundle. The Gysin long exact sequence

$$\cdots \to H^i(Q) \to H^i(Q') \to H^{i-1}(Q) \to H^{i+1}(Q) \to \cdots$$

exists by Remark 2.4.1, and is functorial for maps of the quotient space.

On the other hand, we have the universal principal G-bundle Spec $k \to BG$, whose associated $\mathbb{G}_m$-torsor is the 1-morphism $BN \to BG$. Assuming that G is reductive, we will construct a long exact sequence for this $\mathbb{G}_m$-torsor:

$$\cdots \to H^i(BG) \to H^i(BN) \to H^{i-1}(BG) \to H^{i+1}(BG) \to \cdots$$

In order to do this, we observe that by Lemma 2.3.5, given any integer $s > 0$, there is a principal G-bundle $P \to Q$ with $H^i(P) = 0$ for $i \leq 2s + 2$, so that the classifying morphism $Q \to BG$ is $(2s + 2)$-acyclic. Similarly, if $Q' \to Q$ is the associated $\mathbb{G}_m$-bundle, then $P \to Q'$ is a principal N-bundle, whose classifying morphism $Q' \to BN$ is $(2s+2)$-acyclic. We define the Gysin sequence for $BN \to BG$ by requiring each of the squares in the following diagram to commute, where the vertical isomorphisms are induced by the classifying morphisms for $i \leq 2s$:

$$\begin{array}{ccccccccccc}
\cdots \longrightarrow & H^i(BG) & \longrightarrow & H^i(BN) & \longrightarrow & H^{i-1}(BG) & \longrightarrow & H^{i+1}(BG) & \longrightarrow \cdots \\
& \cong\downarrow & & \cong\downarrow & & \cong\downarrow & & \cong\downarrow & \\
\cdots \longrightarrow & H^i(Q) & \longrightarrow & H^i(Q') & \longrightarrow & H^{i-1}(Q) & \longrightarrow & H^{i+1}(Q) & \longrightarrow \cdots
\end{array}$$

That the Gysin sequence does not depend on the particular principal bundle chosen, but any principal G-bundle $P \to Q$ of the type mentioned in the statement of Lemma 2.3.5 defines the same Gysin sequence for $BN \to BG$, follows from the "independence of (V, S) argument", which was used in the proof of Proposition 2.3.1. More generally, by the independence of (V, S) argument and Lemma 2.3.4, the Gysin sequence for $BN \to BG$ is compatible with the Gysin sequence for principal G-bundles over schemes (even when the vertical morphisms are not isomorphisms) by the same independence of (V, S) argument. Therefore we have proved the following proposition.

Proposition 2.4.3. *If G is reductive, there is a Gysin long exact sequence for the $\mathbb{G}_m$-torsor $BN \to BG$, which is compatible with Gysin sequence for principal G-bundles over schemes in the sense that if $P \to Q$ is a principal G-bundle with Q a smooth, quasi-projective scheme over k, and $Q' \to Q$ the associated $\mathbb{G}_m$-bundle, then all the squares in the following diagram are commutative, where the vertical maps are induced by the classifying morphisms.*

$$\begin{array}{ccccccccccc}
\cdots \longrightarrow & H^i(BG) & \longrightarrow & H^i(BN) & \longrightarrow & H^{i-1}(BG) & \longrightarrow & H^{i+1}(BG) & \longrightarrow \cdots \\
& \downarrow & & \downarrow & & \downarrow & & \downarrow & \\
\cdots \longrightarrow & H^i(Q) & \longrightarrow & H^i(Q') & \longrightarrow & H^{i-1}(Q) & \longrightarrow & H^{i+1}(Q) & \longrightarrow \cdots
\end{array}$$

Remark 2.4.4. The above proposition and its proof are valid for coefficients $\mathbb{Z}/n\mathbb{Z}$ where n is not divisible by the characteristic of the ground field k, with the obvious modification by the introduction of an appropriate Tate twist in each degree.

2.5 Cohomology calculation

In the paper of Holla and Nitsure [H-N-1], the main ingredient of the determination of the singular cohomology ring $H^*_{sing}(BGO(2n); \mathbb{F}_2)$ was the Gysin sequence. Other conceptual points that were used their argument are the following :
(1) The fact that $H^*_{sing}(BO(n); \mathbb{F}_2) \cong \mathbb{F}_2[w_1, \dots, w_n]$, where w_i is the i-th Stiefel-Whitney class;
(2) the splitting principle for $O(n)$-bundles;
(3) the fact that under the map in cohomology induced by the natural morphism

$BO(n) \to BGL_n$ we have $c_i \mapsto w_i^2$, and
(4) the Künneth formula, i.e. that the natural map $H^*_{sing}(\mathbb{C}^\times;\mathbb{F}_2)\otimes H^*_{sing}(X;\mathbb{F}_2) \to H^*_{sing}(\mathbb{C}^\times \times X;\mathbb{F}_2)$ is an isomorphism.

The Gysin sequence for $BO(2n) \to BGO(2n)$ in étale cohomology is already established in Proposition 2.4.3. Among the other conceptual points, an étale cohomology version of (1) is available in [J-1]. (2) The splitting principle in étale cohomology is the 'Claim' at page 180 of [E-K-V], while an étale cohomology version of (3) appears in the same paper. By the Künneth formula in étale cohomology (see Corollaire 1.11 on page 236 of Deligne [SGA $4\frac{1}{2}$]), the natural map $H^*_{et}(X;\mathbb{F}_2) \otimes H^*_{et}(\mathbb{G}_m;\mathbb{F}_2) \to H^*_{et}(X \times \mathbb{G}_m;\mathbb{F}_2)$ is an isomorphism for a finite type scheme X over k.

With this preparation, we observe that the cohomology classes in $H^*_{\text{sing}}(BGO(2n);\mathbb{F}_2)$ introduced in section 4.3 of Holla and Nitsure [H-N-1] make sense in the étale setup, and the proof of Theorem 4.3.5 of [H-N-1] carries over word by word. Therefore the description in terms of explicit generators and relations of the étale cohomology ring $H^*_{et}(BGO(2n);\mathbb{F}_2)$ of the classifying stack of $GO(2n)$ over any algebraically closed field is exactly the same as the reduced mod-2 singular cohomology ring of the topological space $BGO(2n)$, where $GO(2n)$ is considered as a Lie group over complex numbers. Also, the map induced in étale cohomology by the natural morphism $BGO(2n) \to BGL_{2n}$ has exactly the same description as in Proposition 3.2 of [H-N-2].

Characteristic Classes for $SO(2n+1)$

It is known that $H^*(BSO(2n+1)) \cong \mathbb{F}_2[w_2,\dots,w_{2n+1}]$. Nonetheless, for the lack of a suitable reference in the context of algebraic stacks, we sketch a proof of this. Note that $O(2n+1) \cong \mu_2 \times SO(2n+1)$, for $n \geq 1$. Hence, $BO(2n+1) \cong B\mu_2 \times BSO(2n+1)$ for $n \geq 1$. We will show that in this specific case Künneth formula holds for the cohomology of algebraic stacks: $H^*(B\mu_2 \times BSO(2n+1)) \cong H^*(B\mu_2) \otimes H^*(BSO(2n+1))$. This will imply that $\mathbb{F}_2[w_1,\dots,w_{2n+1}] \cong \mathbb{F}_2[w_1] \otimes H^*(BSO(2n+1))$, from which it can be seen that $H^*(BSO(2n+1)) \cong \mathbb{F}_2[w_2,\dots,w_{2n+1}]$.

By the Künneth formula for étale cohomology (see Corollaire 1.11 on page 236 of Deligne [SGA $4\frac{1}{2}$]), if X is any finite type, scheme over k, and if B_r stands for the bundle quotient of $\mathbb{A}^{r+1} - \{0\}$ by $\mu_2 \subset \mathbb{G}_m$, then the natural map $H^*(B_r) \otimes H^*(X) \to H^*(B_r \times X)$ is an isomorphism.

The Künneth formula for the product $B\mu_2 \times BSO(2n+1)$ can now be deduced as follows. As we have described in Section 2.3, we can use acyclic covers to approximate the cohomology of classifying stacks. In the formula $H^*(B_r) \otimes H^*(X) \cong H^*(B_r \times X)$, we substitute for X an acyclic cover for $BSO(2n+1)$, and note that B_r is an r-acyclic cover for $B\mu_2$. This completes the proof.

Chapter 3

Quadric invariants and degeneration in smooth-étale cohomology

Joint work with Nitin Nitsure (arXiv:1302.0092v2)

3.1 Introduction

This paper extends to the algebraic category the topological results of Holla and Nitsure [H-N-1] and [H-N-2]. We address the following three questions, which were answered in the earlier papers in the topological category.

Let S be a scheme on which 2 is invertible, for example, $S = \mathrm{Spec}\,\mathbb{Z}[1/2]$. Let a nondegenerate quadratic triple $T = (E, L, b)$ of rank n on a scheme X over S consist of a rank n vector bundle E and a line bundle L on X, and a nondegenerate symmetric bilinear form b on E taking values in L. As we explain later, such a triple T is the same as a principal $GO_{n,S}$-bundle where $GO_{n,S}$ is the general orthogonal group-scheme over S. Let $BGO_{n,S}$ be its classifying algebraic stack, and let $H^*(BGO_{n,S}, \mathbb{F}_2)$ denote its cohomology ring in the smooth-étale topology. A characteristic class of degree i for nondegenerate quadratic triples of rank n is an element $\alpha \in H^i(BGO_{n,S}, \mathbb{F}_2)$. Such an α associates to any such T an étale cohomology class $\alpha(T) \in H^i(X, \mathbb{F}_2)$, which is functorial in the sense that it commutes with pull-backs under any $X' \to X$.

Question 1 What is the ring $H^*(BGO_{n,S}, \mathbb{F}_2)$ of characteristic classes for nondegenerate quadratic triples of rank n on schemes over S?

A nondegenerate quadric bundle Q on X of rank n will mean an equivalence class of rank n nondegenerate quadratic triples on X, where we regard a triple

$T = (E, L, b)$ as being equivalent to another triple $T \otimes K = (E \otimes K, L \otimes K^2, b \otimes 1_K)$, where K is any line bundle on X (note that the vanishing of the quadratic forms associated to b and $b \otimes 1_K$ define the same closed subscheme $C \subset P(E) = P(E \otimes K)$, which a bundle of nondegenerate quadrics). A cohomological invariant of degree i associates to any such Q an étale cohomology class $\alpha(Q) \in H^i(X, \mathbb{F}_2)$, which is functorial under pull-backs. These form a subring $PH^*(BGO_{n,S}, \mathbb{F}_2) \subset H^*(BGO_{n,S}, \mathbb{F}_2)$.

Question 2 What is the ring $PH^*(BGO_{n,S}, \mathbb{F}_2)$ of cohomological invariants for nondegenerate quadric bundles of rank n on schemes over S?

We will now come to the third question, answering which is the main purpose of this paper. Let X be a regular scheme over $\mathbb{Z}[1/2]$, and let $Y \subset X$ be a closed regular subscheme of pure codimension 1. We are interested in quadric bundles over X which are nondegenerate over $X - Y$ and minimally degenerate over Y. This means that if Q is represented by a quadratic triple $T = (E, L, b)$, then b is nondegenerate of rank n on $X - Y$, while rank of $b|_Y$ is $n - 1$.

Question 3 How do the cohomological invariants of the nondegenerate quadric bundle $Q|_{X-Y}$ behave under degeneration?

Before we come to describing our solution to the question 3, we make some comments on questions 1 and 2. Let $QuasiProj_S$ be the category of all quasi-projective schemes over S. For any group scheme G over S, we have a functor $|BG|$ on $QuasiProj_S$, which associates to any X the set of all isomorphism classes of étale locally trivial G-torsors on X. Let $H^*(-, \mathbb{F}_2)$ be the functor on $QuasiProj_S$ which associates to any X the étale cohomology $H^*(X, \mathbb{F}_2)$. We show in section 4 (Proposition 3.4.1) that when the base S is an excellent Dedekind domain, for any reductive group-scheme G over S, any cohomology class $\alpha \in H^*(BG, \mathbb{F}_2)$ is the same as a morphism of functors α' from $|BG|$ to $H^*(-, \mathbb{F}_2)$ on the category $QuasiProj_S$. Bhaumik [Bh] has recently shown this when the base S is of the form $S = \operatorname{Spec} k$ where k is a field of characteristic $\neq 2$. All that remains for us to do is to point out how the arguments in [Bh] – which were based on an idea of Totaro [T] for approximating classifying stacks BG using Geometric Invariant Theory and Jouanolou's trick – extend from base k to base S, using the extension of GIT to more general bases due to Seshadri [Se].

Our second comment on questions (1) and (2) is as follows. The singular cohomology rings R_n of the CW complexes $BGO_n(\mathbb{C})$ for the complex Lie groups $GO_n(\mathbb{C})$ with $\mathbb{F}_2$ coefficients have been explicitly determined in Holla and Nitsure [H-N-1] and [H-N-2], in terms of generators and relations. The primitive subrings $PR_n \subset R_n$ have been determined in [H-N-2] in a closed form for all odd n, for $n = 4$, and for all n of the form $4m+2$. As shown in [Bh], for any separably closed base field k of characteristic 2, we get the same answers: $H^*(BGO_{n,k}, \mathbb{F}_2) = R_n$ and $PH^*(BGO_{n,k}, \mathbb{F}_2) = PR_n$. The situation over a general base, such as $S = \operatorname{Spec} \mathbb{Z}[1/2]$, is more complicated (see Remark 3.5.15), and will not be described here.

The answer to the question (3) is the main theorem of this paper (Theorem 3.5.1), which describes how the étale cohomological invariants of the nondegenerate quadric bundle over $X - Y$ behave under degeneration. Concretely, we describe how the quadric invariants behave under the Gysin boundary map $\partial : H^*(X - Y, \mathbb{F}_2) \to H^{*-1}(Y, \mathbb{F}_2)$ from the étale cohomology of $X - Y$ to the étale cohomology of the degeneration locus Y.

An analogous result has been proved in [H-N-2] over complex numbers in the topological category using singular cohomology. Some of the crucial techniques used in the topological case – namely, tubular neighbourhoods, orthogonal projections using Riemannian metrics and usual homotopy – do not extend to the algebraic category. We replace the older arguments with the use of algebraic stacks which leads to a much more natural argument – one that also sheds light on the earlier topological result. A by-product of the proof is that the main theorem works also for degenerating quadric bundles defined over regular pairs of algebraic stacks over S, so we have stated and proved it in this generality.

Extensive concrete calculations for the rings R_n and PR_n have been given in [H-N-1] and [H-N-2], and the images of the generators of PR_n under the Gysin boundary map have been listed in all low ranks (up to rank 6) in [H-N-2]. As explained in section 5.5, these calculations now extend to the algebraic category over separably closed fields of characteristic $\neq 2$. We expect this to be of some interest, given the important role that quadratic forms play in arithmetic and algebraic geometry.

This paper is arranged as follows. In section 2, we recall basic facts about absolute purity over algebraic stacks, and prove a 'multiplicity lemma' (Lemma 3.2.5) for Gysin boundary maps which is useful later. The section 3 introduces the basic notions about quadratic triples, quadric bundles and their degenerations, following [H-N-1] and [H-N-2]. We introduce the étale cohomological invariants of nondegenerate quadratic triples and quadric bundles in section 4, and explain why the arguments in [Bh] extend from base a field k to an excellent Dedekind domain S, so as to show that the functorial characteristic classes and cohomological invariants on quasi-projective schemes over such a base S exactly form the rings $H^*(BGO_{n,S}, \mathbb{F}_2)$ and $PH^*(BGO_{n,S}, \mathbb{F}_2)$. The section 5 is devoted to the proof of our main theorem (Theorem 3.5.1). The idea of the proof is to introduce a universal stack $\mathcal{M}$ for at most mildly degenerate quadratic triples (section 5.2) and to simulate a tubular neighbourhood of the degeneration locus $\mathcal{D}$ in $\mathcal{M}$, up to $\mathbf{A}^1$-homotopy, by another algebraic stack (introduced in section 5.3) which is the stackification of the 'standard model' for degenerating quadrics which was earlier introduced in [H-N-2].

3.2 Purity and Gysin boundary maps

Regular pairs of schemes

A **regular scheme** will as usual mean a locally noetherian scheme all whose local rings are regular. We will work over a regular base scheme S over $\mathbb{Z}[1/2]$. All schemes, algebraic spaces and algebraic stacks and all morphisms will be assumed to be over S, and the structure morphisms to S will be assumed to be quasi-compact. Important extreme examples are when $S = \operatorname{Spec} \mathbb{Z}[1/2]$ or when $S = \operatorname{Spec} k$ for an algebraically closed field of characteristic $\neq 2$.

A **regular pair** (X, Y) **of schemes of pure codimension** 1 (or simply a 'regular pair', as the codimension will always be assumed to be 1 in what follows) consists of a regular scheme X and a closed subscheme Y which is regular and of codimension 1 in X at all points of Y. In particular, Y is an effective Cartier divisor in X. A special class of regular pairs are the **smooth pairs over the base** S, where S is a given regular scheme, and X and Y are smooth S-schemes with Y a closed subscheme of X, which we will assume to have codimension 1. An example of a regular pair is where $X = \operatorname{Spec} \mathbb{Z}[1/2]$ and $Y = \operatorname{Spec} \mathbb{F}_p$ where p is an odd prime. Note that in this example X is smooth over the base $S = \operatorname{Spec} \mathbb{Z}[1/2]$ but Y is not, so this regular pair is not a smooth pair over S. Regular or smooth pairs of algebraic spaces have a similar definition.

Recall that if (X, Y) is a regular pair of algebraic spaces of codimension 1 such that 2 is invertible on X, then the absolute purity theorem of Gabber (originally conjectured by Grothendieck – see [Fu] for Gabber's proof) gives us the following.

3.2.1 Gabber's absolute purity theorem. Under the above hypothesis, the local cohomology sheaves $\underline{H}^r_Y(\mathbb{F}_{2,X})$ on X are given by

$$\underline{H}^r_Y(\mathbb{F}_{2,X}) = \begin{cases} 0 & \text{for } r \neq 2, \\ i_*\mathbb{F}_{2,Y} & \text{for } r = 2. \end{cases}$$

By the resulting degenerateness of the local-global spectral sequence, this gives the natural identifications

$$H^r_Y(X, \mathbb{F}_2) = H^{r-2}(X, \underline{H}^2_Y(\mathbb{F}_{2,X})) = H^{r-2}(Y, \mathbb{F}_2).$$

If moreover Y is irreducible, then under the above isomorphism the **fundamental class** $s_{Y/X} \in H^2_Y(X, \mathbb{F}_2)$ of Y in X corresponds to the generating section in $H^0(X, \underline{H}^2_Y(\mathbb{F}_{2,X}))$.

Under the isomorphism $H^r_Y(X, \mathbb{F}_2) \cong H^{r-2}(Y, \mathbb{F}_2)$, the long exact sequence for cohomology with supports Y becomes the long exact Gysin sequence

$$\ldots \to H^r(X, \mathbb{F}_2) \to H^r(X - Y, \mathbb{F}_2) \stackrel{\partial}{\to} H^{r-1}(Y, \mathbb{F}_2) \to H^{r+1}(X, \mathbb{F}_2) \to \ldots$$

for the pair (X, Y). The maps $\partial : H^r(X - Y, \mathbb{F}_2) \to H^{r-1}(Y, \mathbb{F}_2)$ are the **Gysin boundary maps**.

Cohomology and local cohomology for algebraic stacks

We largely follow the terminology of Laumon and Moret-Bailly [L-MB] for algebraic stacks. Recall that we work over a regular base scheme S over $\mathbb{Z}[1/2]$, and all schemes, algebraic spaces and algebraic stacks are assumed to be quasi-compact over S. The word 'stack' will signify such an algebraic stack, and morphisms will always be over S.

Let *Lis-Et*$(\mathfrak{X})$ denote the smooth-étale site of $\mathfrak{X}$, and let $\mathfrak{X}_{lis\text{-}et}$ be the resulting topos, which we will denote simply by $\mathfrak{X}$. All sheaves and cohomology for $\mathfrak{X}$ will be in the smooth-étale topology. The sheaf of main interest to us is the constant sheaf $\mathbb{F}_{2,\mathfrak{X}}$ in $\mathfrak{X}_{lis\text{-}et}$. This sheaf is Cartesian and representable. We will simply denote it by $\mathbb{F}_2$. We denote by $Mod_{cart}(\mathfrak{X}, \mathbb{F}_2)$ the abelian category of Cartesian sheaves of $\mathbb{F}_2$-modules on $\mathfrak{X}$, and by $D^+_{cart}(\mathfrak{X}, \mathbb{F}_2)$ the derived category of cohomologically Cartesian bounded below complexes of sheaves of $\mathbb{F}_2$-modules on $\mathfrak{X}$. For an algebraic space X, the inclusion functor $Et(X) \hookrightarrow Lis\text{-}Et(X)$ from the étale to the smooth site induces a geometric morphism of topoi $X_{lis\text{-}et} \to X_{et}$ under which we get an equivalence between X_{et} and the subcategory of Cartesian sheaves in $X_{lis\text{-}et}$ (see [L-MB] 12.2.3). This induces an equivalence between $D^+(X_{et}, \mathbb{F}_2)$ and $D^+_{cart}(X_{lis\text{-}et}, \mathbb{F}_2)$. Consequently, étale cohomology (or hypercohomology) on X_{et} where X is an algebraic space may be regarded as the smooth-étale cohomology (or hypercohomology) of the corresponding Cartesian object on $X_{lis\text{-}et}$.

It is known due to Behrend and Gabber that a 1-morphism $f : \mathfrak{X}' \to \mathfrak{X}$ of stacks does not functorially induce a geometric morphism of topoi from $\mathfrak{X}'_{lis\text{-}et}$ to $\mathfrak{X}_{lis\text{-}et}$. However, Olsson showed in [O] how to functorially associate to f a pull-back functor

$$f^* : D^+_{cart}(\mathfrak{X}, \mathbb{F}_2) \to D^+_{cart}(\mathfrak{X}', \mathbb{F}_2)$$

of triangulated categories. The facts about f^* that we need are only a small part of the theory of six operations for algebraic stacks developed by Behrend [Be1], Olsson [O], Laszlo and Olsson [L-O], and other authors.

Notation: The functor $f^* : D^+_{cart}(\mathfrak{X}, \mathbb{F}_2) \to D^+_{cart}(\mathfrak{X}', \mathbb{F}_2)$ is actually called f^{-1} in [O]. Our notation f^* for it follows the section 4.3 of [L-O].

We now briefly recall the construction of f^* from [O]. Let $X \to \mathfrak{X}$ be a smooth atlas where X is an algebraic space over S, and let $X^+_\bullet = cosk_0^+(X/\mathfrak{X})$ be its nerve regarded as a strictly simplicial algebraic space, where X_n is the fiber product of $n+1$ copies of X over $\mathfrak{X}$. The word 'strict' means that only the morphisms $X_n \to X_m$ in $cosk_0(X/\mathfrak{X})$ which correspond to strictly monotonic maps $[m] \to [n]$ (and which are therefore smooth morphisms) are retained as part of the structure of $X^+_\bullet$. As these morphisms are smooth, the topos $X^+_{\bullet,lis\text{-}et}$ makes sense. Let $X^+_{\bullet,et}$ be the corresponding topos with étale topology.

There are natural geometric morphisms of topoi $\pi : X^+_{\bullet,lis\text{-}et} \to \mathfrak{X}_{lis\text{-}et}$ and $\epsilon : X^+_{\bullet,lis\text{-}et} \to X^+_{\bullet,et}$. Using the theory of cohomological descent developed in [SGA 4] and a result of Gabber, Olsson proves (Theorem 4.7 of [O]) that the induced functors $\pi^* : D^+_{cart}(\mathfrak{X}_{lis\text{-}et}) \to D^+_{cart}(X^+_{\bullet,lis\text{-}et})$ and $\epsilon^* : D^+_{cart}(X^+_{\bullet,et}) \to D^+_{cart}(X^+_{\bullet,lis\text{-}et})$ are equivalences of triangulated categories, with quasi-inverses respectively $R\pi_*$ and ϵ_*.

Note: The Cartesian conditions for $\mathfrak{X}_{lis\text{-}et}$ and for $X^+_{\bullet,et}$ are meant respectively in the sheaf sense and in the strictly simplicial sense, while the Cartesian condition for $X^+_{\bullet,lis\text{-}et}$ is meant to be simultaneously satisfied in both the sheaf and the strictly simplicial senses.

The composite $\epsilon_* \circ \pi^*$ defines an equivalence of triangulated categories

$$\psi : D^+_{cart}(\mathfrak{X}, \mathbb{F}_2) \to D^+_{cart}(X^+_{\bullet,et}, \mathbb{F}_2).$$

Given a 1-morphism of S-stacks $f : \mathfrak{X}' \to \mathfrak{X}$, let $\mathfrak{X}'' = \mathfrak{X}' \times_{\mathfrak{X}} X$ (which is a stack), and let $X' \to \mathfrak{X}''$ be a smooth atlas where X' is an algebraic space. Then we can define a similar equivalence of triangulated categories $\psi' : D^+_{cart}(\mathfrak{X}', \mathbb{F}_2) \to D^+_{cart}(X'^+_\bullet, \mathbb{F}_2)$. The induced morphism $f_\bullet : X'^+_\bullet \to X^+_\bullet$ actually gives a geometric morphism of topoi

$$(f^*_\bullet, f_{\bullet,*}) : X'^+_{\bullet,et} \to X^+_{\bullet,et}$$

so it induces a functor of triangulated categories

$$f^*_\bullet : D^+_{cart}(X^+_{\bullet,et}, \mathbb{F}_2) \to D^+_{cart}(X'^+_\bullet, \mathbb{F}_2).$$

Finally, Olsson defines (see 9.16 of [O]) the functor $f^* : D^+_{cart}(\mathfrak{X}, \mathbb{F}_2) \to D^+_{cart}(\mathfrak{X}', \mathbb{F}_2)$ as the composite

$$D^+_{cart}(\mathfrak{X}, \mathbb{F}_2) \overset{\psi}{\to} D^+_{cart}(X^+_{\bullet,et}, \mathbb{F}_2) \overset{f^*_\bullet}{\to} D^+_{cart}(X'^+_{\bullet,et}, \mathbb{F}_2) \overset{\psi'^{-1}}{\to} D^+_{cart}(\mathfrak{X}', \mathbb{F}_2).$$

It is shown in [O] that the above functor f^* is left adjoint to the functor $Rf_* : D^+_{cart}(\mathfrak{X}', \mathbb{F}_2) \to D^+_{cart}(\mathfrak{X}, \mathbb{F}_2)$. Hence it is well-defined, independent of the choices of the smooth atlases $X \to \mathfrak{X}$ and $X' \to \mathfrak{X}''$. Moreover, it is functorial in f, that is, given $\mathfrak{X} \xrightarrow{f} \mathfrak{Y} \xrightarrow{g} \mathfrak{Z}$, we have $(g \circ f)^* = f^* \circ g^*$.

Note In the special case where the 1-morphism $f : \mathfrak{X}' \to \mathfrak{X}$ is smooth (but not necessarily representable), the pullback $f^* : \mathfrak{X}_{lis\text{-}et} \to \mathfrak{X}'_{lis\text{-}et}$ on sheaves of sets is indeed (left) exact, and so $f^* : D^+_{cart}(\mathfrak{X}, \mathbb{F}_2) \to D^+_{cart}(\mathfrak{X}', \mathbb{F}_2)$ can be defined directly. As one would desire, the direct definition of $f^* : D^+_{cart}(\mathfrak{X}, \mathbb{F}_2) \to D^+_{cart}(\mathfrak{X}', \mathbb{F}_2)$ when f is smooth coincides with the above indirect general definition.

For any $f : \mathfrak{X}' \to \mathfrak{X}$, with the above definition of the functor $f^* : D^+_{cart}(\mathfrak{X}, \mathbb{F}_2) \to D^+_{cart}(\mathfrak{X}', \mathbb{F}_2)$, we have $f^*\mathbb{F}_{2,\mathfrak{X}} = \mathbb{F}_{2,\mathfrak{X}'}$. Hence for any F in $D^+_{cart}(\mathfrak{X}, \mathbb{F}_2)$, the functor $f^* : D^+_{cart}(\mathfrak{X}, \mathbb{F}_2) \to D^+_{cart}(\mathfrak{X}', \mathbb{F}_2)$ induces a pull back homomorphism f^* on hypercohomology

$$\mathbb{H}^i(\mathfrak{X}, F) = Hom_{D^+_{cart}(\mathfrak{X},\mathbb{F}_2)}(\mathbb{F}_2, F[i]) \to Hom_{D^+_{cart}(\mathfrak{X}',\mathbb{F}_2)}(\mathbb{F}_2, f^*F[i]) = \mathbb{H}^i(\mathfrak{X}', f^*F).$$

In the important special case where $F = \mathbb{F}_2$, we thus get homomorphisms

$$f^* : H^i(\mathfrak{X}, \mathbb{F}_2) \to H^i(\mathfrak{X}', \mathbb{F}_2).$$

If $i : \mathfrak{Y} \hookrightarrow \mathfrak{X}$ is a closed substack of an algebraic stack $\mathfrak{X}$ over S, and if F is a sheaf in $\mathfrak{X}_{lis\text{-}et}$, then recall that the (0)-th local cohomology sheaf $\underline{H}^0_{\mathfrak{Y}}(F) = i_*i^!(F)$ can be directly defined as follows. For any U in $Lis\text{-}Et(\mathfrak{X})$, let $V \subset U$ be the pullback of $\mathfrak{Y}$. Then $\underline{H}^0_{\mathfrak{Y}}(F)(U)$ is defined to be the kernel of the restriction map $F(U) \to F(U - V)$. The functor $F \mapsto \underline{H}^0_{\mathfrak{Y}}(F)$ is left exact, and so we have its derived functor $i_*Ri^! : D^+_{cart}(\mathfrak{X}, \mathbb{F}_2) \to D^+_{cart}(\mathfrak{X}, \mathbb{F}_2)$. (*Note*: It is enough for our purpose to directly define $i_*Ri^!$, without first defining $Ri^!$ and then composing it with i_*.)

If $j : \mathfrak{X} - \mathfrak{Y} \hookrightarrow \mathfrak{X}$ is the complementary open inclusion, then for any F in $D^+_{cart}(\mathfrak{X}, \mathbb{F}_2)$ we have a functorial exact triangle $(i_*Ri^!\mathbb{F}_{2,\mathfrak{X}} \to \mathbb{F}_{2,\mathfrak{X}} \to j_*j^*\mathbb{F}_{2,\mathfrak{X}} \to)$ in $D^+_{cart}(\mathfrak{X}, \mathbb{F}_2)$, which corresponds under the equivalence $\psi : D^+_{cart}(\mathfrak{X}, \mathbb{F}_2) \to D^+_{cart}(X^+_{\bullet,et}, \mathbb{F}_2)$ to the exact triangle $(i_{\bullet,*}Ri^!_{\bullet}\mathbb{F}_{2,X^+_{\bullet,et}} \to \mathbb{F}_{2,X^+_{\bullet,et}} \to j_{\bullet,*}j^*_{\bullet}\mathbb{F}_{2,X^+_{\bullet,et}} \to)$ in $D^+_{cart}(X^+_{\bullet,et}, \mathbb{F}_2)$. Here, $Y \subset X$ is the pullback of $\mathfrak{Y}$ under the atlas $X \to \mathfrak{X}$, $i_{\bullet} : Y^+_{\bullet} \hookrightarrow X^+_{\bullet}$ is the corresponding closed strictly simplicial subspace and $j_{\bullet} : X^+_{\bullet} - Y^+_{\bullet} \hookrightarrow X^+_{\bullet}$ is the complementary open inclusion.

The local cohomology groups are defined in terms of $i_*Ri^!$ as the hypercohomologies

$$H^r_{\mathfrak{Y}}(\mathfrak{X}, \mathbb{F}_2) = \mathbb{H}^r(\mathfrak{X}, i_*Ri^!\mathbb{F}_2).$$

Let $i : \mathfrak{Y} \hookrightarrow \mathfrak{X}$ (resp. $i' : \mathfrak{Y}' \hookrightarrow \mathfrak{X}'$) be closed substacks, such that the closed substack $f^{-1}(\mathfrak{Y}) \subset \mathfrak{X}'$ has the same support as $\mathfrak{Y}'$. We have a natural map

$$f^*i_*Ri^!\mathbb{F}_{2,\mathfrak{X}} \to i'_*Ri'^*\mathbb{F}_{2,\mathfrak{X}'}$$

in $D^+_{cart}(\mathfrak{X}')$ (see below), which at the level of hypercohomologies gives a pull back homomorphism on local cohomologies

$$f^* : H^i_{\mathfrak{Y}}(\mathfrak{X}, \mathbb{F}_2) \to H^i_{\mathfrak{Y}'}(\mathfrak{X}', \mathbb{F}_2).$$

The homomorphism $f^*i_*Ri^!\mathbb{F}_{2,\mathfrak{X}} \to i'_*Ri'^*\mathbb{F}_{2,\mathfrak{X}'}$ has the following strictly simplicial construction. Choosing atlases as above, let $i_\bullet : Y^+_\bullet \subset X^+_\bullet$ and $i'_\bullet : Y'^+_\bullet \subset X'^+_\bullet$ be the corresponding inverse images. Note that f gives $f_\bullet : X'^+_\bullet \to X^+_\bullet$ under which the support of $f^{-1}_\bullet(Y^+_\bullet)$ is the support of $Y'^+_\bullet$. Then $f_\bullet$ induces a commutative diagram (morphism of exact triangles) in $D^+_{cart}(X'^+_{\bullet,et}, \mathbb{F}_2)$

$$\begin{array}{ccccccc} f^*_\bullet i_{\bullet,*}Ri^!_\bullet\mathbb{F}_{2,X^+_{\bullet,et}} & \to & f^*_\bullet\mathbb{F}_{2,X^+_{\bullet,et}} & \to & f^*_\bullet j_{\bullet,*}j^*_\bullet\mathbb{F}_{2,X^+_{\bullet,et}} & \to \\ \downarrow & & \downarrow & & \downarrow & \\ i'_{\bullet,*}Ri'^!_\bullet\mathbb{F}_{2,X'^+_{\bullet,et}} & \to & \mathbb{F}_{2,X'^+_{\bullet,et}} & \to & j'_{\bullet,*}j'^*_\bullet\mathbb{F}_{2,X'^+_{\bullet,et}} & \to \end{array}$$

Hence we get the following commutative diagram in $D^+_{cart}(\mathfrak{X}', \mathbb{F}_2)$

$$\begin{array}{ccccccc} f^*i_*Ri^!\mathbb{F}_{2,\mathfrak{X}} & \to & f^*\mathbb{F}_{2,\mathfrak{X}} & \to & f^*j_*j^*\mathbb{F}_{2,\mathfrak{X}} & \to \\ \downarrow & & \downarrow & & \downarrow & \\ i'_*Ri'^!\mathbb{F}_{2,\mathfrak{X}'} & \to & \mathbb{F}_{2,\mathfrak{X}'} & \to & j'_*j'^*\mathbb{F}_{2,\mathfrak{X}'} & \to \end{array}$$

Remark 3.2.2 To check that a homomorphism in $D^+_{cart}(\mathfrak{X}, \mathbb{F}_2)$ is zero (or to check that a diagram commutes in $D^+_{cart}(\mathfrak{X}, \mathbb{F}_2)$), it is *not enough* to check that its pull back is zero in $D^+(X_{et}, \mathbb{F}_2)$ for an atlas $X \to \mathfrak{X}$ – rather, one needs to check that it is zero in the derived category of the strictly simplicial topos $X^+_{\bullet,et}$, so that cohomological descent can be employed. However, if it is a homomorphism (or a diagram) coming from $Mod_{cart}(\mathfrak{X}, \mathbb{F}_2)$, to check that it is zero (or commutes) in $D^+_{cart}(\mathfrak{X}, \mathbb{F}_2)$, it is enough to check that its pull back to $Mod(X_{et}, \mathbb{F}_2)$ is zero (or commutes).

Remark 3.2.3 By taking hypercohomologies, the above morphism of exact triangles gives a commutative diagram

$$\begin{array}{ccccccc} H^i(\mathfrak{X}, \mathbb{F}_2) & \to & H^i(\mathfrak{X} - \mathfrak{Y}, \mathbb{F}_2) & \stackrel{d}{\to} & H^{i+1}_{\mathfrak{Y}}(\mathfrak{X}, \mathbb{F}_2) & \to & H^{i+1}(\mathfrak{X}, \mathbb{F}_2) \\ \downarrow & & \downarrow & & \downarrow & & \downarrow \\ H^i(\mathfrak{X}', \mathbb{F}_2) & \to & H^i(\mathfrak{X}' - \mathfrak{Y}', \mathbb{F}_2) & \stackrel{d}{\to} & H^{i+1}_{\mathfrak{Y}'}(\mathfrak{X}', \mathbb{F}_2) & \to & H^{i+1}(\mathfrak{X}', \mathbb{F}_2) \end{array}$$

Regular pairs of algebraic stacks

Recall that an algebraic stack $\mathfrak{X}$ is said to be a **regular stack** if for some (hence for every) smooth atlas $X \to \mathfrak{X}$ where X is a scheme (or an algebraic space), the total space X is regular. A **regular pair** $(\mathfrak{X},\mathfrak{Y})$ **of stacks** will mean a closed embedding $i:\mathfrak{Y}\hookrightarrow\mathfrak{X}$ such that both $\mathfrak{X}$ and $\mathfrak{Y}$ are regular stacks, and $\mathfrak{Y}$ has pure codimension 1. Certain crucial examples of such pairs for us, namely the pairs $(\mathcal{M},\mathcal{D})$ and $(\mathcal{N},\mathcal{Z})$ that we introduce later, are in fact **smooth pairs of S-stacks**, that is, the base scheme S is regular and the structure morphisms $\mathfrak{X}\to S$ and $\mathfrak{Y}\to S$ are smooth.

3.2.4 (Absolute purity theorem for algebraic stacks.) With all the above preparation, it follows from the absolute purity theorem for schemes that the absolute purity theorem (Statement 3.2.1) holds also for regular pairs of stacks $(\mathfrak{X},\mathfrak{Y})$ of pure codimension 1 (see [L-O] section 4.10). In particular, we get a long exact Gysin sequence as in Statement 3.2.1.

Lemma 3.2.5. *Let $(\mathfrak{X},\mathfrak{Y})$ and $(\mathfrak{X}',\mathfrak{Y}')$ be regular pairs of algebraic stacks of pure codimension 1 such that $\mathfrak{Y}$ and $\mathfrak{Y}'$ are connected, and 2 is invertible on $\mathfrak{X}$ and $\mathfrak{X}'$. Let $f:\mathfrak{X}'\to\mathfrak{X}$ be a 1-morphism such that the closed substack $\mathfrak{Y}''=\mathfrak{X}'\times_{\mathfrak{X}}\mathfrak{Y}$ of $\mathfrak{Y}$ is equal to $m\mathfrak{Y}'$ for some integer $m\geq 1$ (that is, its ideal sheaf $I_{\mathfrak{Y}''}$ equals $I^m_{\mathfrak{Y}'}$). Then the following statements hold.*
(1) Under $f^*: H^2_{\mathfrak{Y}}(\mathfrak{X},\mathbb{F}_2)\to H^2_{\mathfrak{Y}'}(\mathfrak{X}',\mathbb{F}_2)$, the fundamental classes $s_{\mathfrak{Y}/\mathfrak{X}}$ and $s_{\mathfrak{Y}'/\mathfrak{X}'}$ are related by $s_{\mathfrak{Y}/\mathfrak{X}}\mapsto m\cdot s_{\mathfrak{Y}'/\mathfrak{X}'}$.
(2) The following diagram commutes

$$\begin{array}{ccccccc}
H^i(\mathfrak{X},\mathbb{F}_2) & \to & H^i(\mathfrak{X}-\mathfrak{Y},\mathbb{F}_2) & \stackrel{\partial}{\to} & H^{i-1}(\mathfrak{Y},\mathbb{F}_2) & \to & H^{i+1}(\mathfrak{X},\mathbb{F}_2)\\
\downarrow f^* & & \downarrow f^*_{\mathfrak{X}'-\mathfrak{Y}'} & & \downarrow mf^*_{\mathfrak{Y}'} & & \downarrow f^*\\
H^i(\mathfrak{X}',\mathbb{F}_2) & \to & H^i(\mathfrak{X}'-\mathfrak{Y}',\mathbb{F}_2) & \stackrel{\partial'}{\to} & H^{i-1}(\mathfrak{Y}',\mathbb{F}_2) & \to & H^{i+1}(\mathfrak{X}',\mathbb{F}_2)
\end{array}$$

where the rows are the respective long exact Gysin sequences. In particular,

$$\partial\circ f^*_{\mathfrak{X}'-\mathfrak{Y}'}=m\cdot f^*_{\mathfrak{Y}}\circ\partial'.$$

Proof The statement (1) for smooth pairs of schemes is due to Grothendieck (see [SGA $4\frac{1}{2}$] page 138), who used the purity theorem for smooth pairs. If we instead use the absolute purity theorem, the same proof applies to prove the statement (1) for regular pairs (X,Y) of schemes or algebraic spaces.

Now consider the special case when (X,Y) and (X',Y') are regular pairs of algebraic spaces, and $f:X'\to X$ is a morphism for which $f^{-1}(Y)=mY'$. As $s_{Y/X}\mapsto m\cdot s_{Y'/X'}$, the following diagram commutes, where the horizontal isomorphisms come from purity, and where the right hand vertical map is multiplication by m.

$$\begin{array}{ccc} f^* i_* Ri^! \mathbb{F}_{2,X} & \tilde{\rightarrow} & f^* i_* \mathbb{F}_{2,Y}[-2] \\ \downarrow & & \downarrow m \\ i'_* Ri'^! \mathbb{F}_{2,X'} & \tilde{\rightarrow} & i_* \mathbb{F}_{2,Y'}[-2] \end{array}$$

Note that the above is actually (the shift by -2 of) a commutative diagram coming from $Mod_{cart}(X', \mathbb{F}_2)$. Hence it follows by Remark 3.2.2 that the following diagram commutes in $D^+_{cart}(\mathfrak{X}', \mathbb{F}_2)$.

$$\begin{array}{ccc} f^* i_* Ri^! \mathbb{F}_{2,\mathfrak{X}} & \tilde{\rightarrow} & f^* i_* \mathbb{F}_{2,\mathfrak{Y}}[-2] \\ \downarrow & & \downarrow m \\ i'_* Ri'^! \mathbb{F}_{2,\mathfrak{X}'} & \tilde{\rightarrow} & i_* \mathbb{F}_{2,\mathfrak{Y}'}[-2] \end{array}$$

The statement (1) follows for stacks by applying hypercohomology functors $\mathbb{H}^2$ to the above diagram and chasing the image of the fundamental class $s_{\mathfrak{Y}/\mathfrak{X}}$ in the resulting diagram.

Applying $(i+1)$-th hypercohomology functors to the above diagram, it follows that the following diagram commutes.

$$\begin{array}{ccc} H^{i+1}_{\mathfrak{Y}}(\mathfrak{X}, \mathbb{F}_2) & \tilde{\rightarrow} & H^{i-1}(\mathfrak{Y}, \mathbb{F}_2) \\ f^* \downarrow & & \downarrow m\, f^*_{\mathfrak{Y}'} \\ H^{i+1}_{\mathfrak{Y}'}(\mathfrak{X}', \mathbb{F}_2) & \tilde{\rightarrow} & H^{i-1}(\mathfrak{Y}', \mathbb{F}_2) \end{array}$$

By combining the above commutative diagram with the one given by Remark 3.2.3, the statement (2) follows. □

Remark 3.2.6 The conclusions of Lemma 3.2.5.(2) remains true even if in the hypothesis of the lemma the stacks Y and Y' are not assumed to be connected and the multiplicity m is allowed to vary as a locally constant function over Y'. To see this, note that if Y_i are the connected components of Y and Y'_j are the connected components of Y', and if Y'_{j0} maps into Y_{i_0}, then we can apply the lemma to the pairs $(X - \cup_{i \neq i_0} Y_i, Y_{i_0})$ and $(X' - \cup_{j \neq j_0} Y'_j, Y'_{j_0})$, and conclude by excision.

Remark 3.2.7 Any smooth atlas $X \to \mathfrak{X}$ of an S-stack functorially gives a smooth atlas $X \times \mathbf{A}^1 \to \mathfrak{X} \times \mathbf{A}^1$ where $\mathbf{A}^1 = \operatorname{Spec} \mathbb{Z}[t]$. Using these atlases and the corresponding property for algebraic spaces, it follows from the definition of the pullback map on cohomologies of stacks that if $f, g \in \Gamma(\mathfrak{X}, \mathcal{O}_{\mathfrak{X}})$ are regular functions with graphs $\gamma_f, \gamma_g : \mathfrak{X} \to \mathfrak{X} \times \mathbf{A}^1$, then we have $\gamma_f^* = \gamma_g^* : H^*(\mathfrak{X} \times \mathbf{A}^1, \mathbb{F}_2) \to H^*(\mathfrak{X}, \mathbb{F}_2)$.

3.3 Quadric bundles and their degenerations

Quadratic triples and quadric bundles

Let S be a scheme over $\mathbb{Z}[1/2]$. A **quadratic triple** $T = (E, L, b)$ on an S-scheme X by definition consists of a bilinear form b on a vector bundle E on X, taking values in a line bundle L on X, such that b is nowhere zero on X, that is, b_x is not identically zero on any fiber E_x. We call $\mathrm{rank}(E)$ (which we assume to be globally constant) the **dimension** of the triple T and the $\mathrm{rank}(b_x)$ of the quadratic form b_x on the fiber $E_x = k(x) \otimes_{\mathcal{O}_X} E$ as the **rank** of T at $x \in X$. We will say that a quadratic triple $T = (E, L, b)$ on X is **nondegenerate** if at each $x \in X$ we have $\mathrm{rank}(b_x) = \mathrm{rank}(E)$, equivalently, the $\mathcal{O}_X$-linear map $b : E \to L \otimes_{\mathcal{O}_X} E^*$ is an isomorphism.

Given a triple $T = (E, L, b)$ and a line bundle K on X, the triple $T \otimes K$ is defined to be $(E \otimes K, L \otimes K^{\otimes 2}, b \otimes 1_K)$. We say that two triples $T = (E, L, b)$ and $T' = (E', L', b')$ on X are **equivalent triples** if there exists a line bundle K on X such that $T \otimes K$ is isomorphic to T'. A **quadric bundle** Q over X is by definition an equivalence class $[E, L, b]$ of triples such that for each $x \in X$ we have $b_x \neq 0$. As b_x is not identically zero on any fiber E_x by assumption, we see that $[E, L, b]$ defines a closed subscheme $C \subset P(E) = \mathrm{Proj}\,\mathrm{Sym}(E^*)$ by the vanishing of the quadratic form corresponding to b, which is flat over X and whose each fiber is a quadric hypersurface $C_x \subset P(E_x)$. Conversely the equivalence class $Q = [E, L, b]$ can be recovered from a closed subscheme $C \subset P(E)$ in a banal projective bundle, such that C is flat over X and every schematic fiber is a quadric hypersurface.

Quadratic triples and quadric bundles over algebraic stacks: In the above definitions, the base space for quadratic triples and quadric bundles was taken to be an S-scheme X. But the same definitions work equally well when X is an algebraic stack over S. In what follows, the base spaces for our quadratic triples or quadric bundles will be assumed to be algebraic stacks over S, unless otherwise indicated.

Minimally degenerate quadric bundles, associated nondegenerate triples

The **discriminant** $\det(T)$ of a quadratic triple $T = (E, L, b)$ on X is defined by

$$\det(T) = \det(b) \in \Gamma(X, L^{\mathrm{rank}(E)} \otimes \det(E)^{-2}).$$

A quadratic triple T is nondegenerate if and only if $\det(T)$ is nowhere vanishing. Note that

$$\det(T) = \det(T \otimes K)$$

for any line bundle K on X, so for any quadric bundle $Q = [E, L, b]$. This enables us to define **the discriminant** $\det(Q)$ **of a quadric bundle** to be $\det(T)$ for any representative T.

A quadratic triple (E, L, b) on base Y is called a **minimally degenerate triple** if (i) $\det(b) = 0 \in \Gamma(Y, L^{\mathrm{rank}(E)} \otimes \det(E)^{-2})$ and (ii) $\mathrm{rank}(b_y) = n - 1$ for all $y \in Y$. (Note that if Y is reduced then (ii) implies (i), but not otherwise.) The equivalence class $Q = [E, L, b]$ of such a triple will be called a **minimally degenerate quadric bundle** (this is well defined, independent of the choice of a representative (E, L, b) of Q). If we regard b as an $\mathcal{O}_Y$-linear homomorphism $b : E \to L \otimes E^*$, then $\ker(b)$ is a line bundle, the quotient $\overline{E} = E/\ker(b)$ is a vector bundle of rank equal to $\mathrm{rank}(E) - 1$, and b induces an L-valued quadratic form $\bar{b}$ on $\overline{E}$, giving a nondegenerate quadratic triple $(\overline{E}, L, \bar{b})$.

The next lemmas follows from a simple computation which we omit.

Lemma 3.3.1. *If $Q = [E, L, b]$ is a minimally degenerate quadric bundle on Y, then the nondegenerate quadratic triple*

$$\mathbb{T}^Q = (\overline{E}, L, \bar{b}) \otimes \ker(b)^{-1}$$

is well defined, that is, it remains unaltered if the representative (E, L, b) is replaced by $(E, L, b) \otimes K$ for any line bundle K on Y. □

We will call the above quadratic triple $\mathbb{T}^Q = (\overline{E}, L, \bar{b}) \otimes \ker(b)^{-1}$ (which was originally introduced in [H-N-2]) the **associated nondegenerate quadratic triple** to the minimally degenerate quadric bundle Q.

Mildly degenerating triples on a regular pair (X, Y)

A **mildly degenerating triple** (E, L, b) on a regular pair (X, Y) of stacks is a quadratic triple $T = (E, L, b)$ on X such that $T|_{X-Y}$ is nondegenerate while $T|_Y$ is minimally degenerate, that is, $\mathrm{rank}_x(b) = \mathrm{rank}(E)$ for each $x \in X - Y$ while $\mathrm{rank}_y(b) = \mathrm{rank}(E) - 1$ for each $y \in Y$. As such a Y is reduced, the discriminant $\det(T)$ vanishes on Y.

Given a mildly degenerating triple $T = (E, L, b)$ on a regular pair (X, Y) of schemes, the vanishing multiplicity of the discriminant $\det(T)$ along the divisor Y defines a function $\nu_Y(T) : |Y| \to \mathbb{Z}_{\geq 0}$ on the set $|Y|$ of connected components of Y. This function will be called the **degeneration multiplicity** of the triple. If (X, Y) is a regular pair and $m : |Y| \to \mathbb{Z}_{\geq 0}$ is a continuous function, then we get a new ideal sheaf $I_Y^m \subset \mathcal{O}_X$, and we denote by $mY \subset X$ the corresponding closed subscheme. Note that if $m = 0$ then $mY = \emptyset$ and if $m = 1$ then $mY = Y$. The above has an obvious interpretation also for mildly degenerating triple on a regular pair of stacks.

Remark 3.3.2 In particular, if (X, Y) is a regular pair and if Q is a mildly degenerating quadric bundle on (X, Y), then the restriction Q_Y is a minimally degenerate quadric bundle on Y, which gives us an associated nondegenerate quadratic triple $\mathbb{T}^{Q_Y}$.

3.4 Cohomological invariants

The group schemes GO_n and stacks BGO_n

The **general orthogonal group-scheme** GO_n over $\mathbb{Z}[1/2]$ is the closed subgroup scheme of GL_n over $\mathbb{Z}[1/2]$, whose R-valued points for any $\mathbb{Z}[1/2]$-algebra R (that is, a commutative ring R in which 2 is invertible) are all the $n \times n$ matrices g over R such that ${}^tgg = \lambda I$ for some invertible element $\lambda \in R^\times$. We have a surjective homomorphism $\sigma : GO_n \to \mathbb{G}_m$ of group schemes defined on valued points by $g \mapsto \lambda$ where ${}^tgg = \lambda I$. This has kernel O_n, the orthogonal group scheme over $\mathbb{Z}[1/2]$. As both O_n and $\mathbb{G}_m$ are smooth over $\mathbb{Z}[1/2]$, it follows that GO_n is also smooth over $\mathbb{Z}[1/2]$. In fact, it can be seen that GO_n is a reductive group scheme over $\mathbb{Z}[1/2]$.

For any scheme S over $\mathbb{Z}[1/2]$, the group-scheme $GO_{n,S}$ is obtained by base change, and has a similar description in terms of R-valued points for affine schemes $\operatorname{Spec} R \to S$. The group schemes $GL_{n,S}$, $\mathbb{G}_{m,S}$, $GO_{n,S}$, $O_{n,S}$, etc. will be simply denoted by GL_n, $\mathbb{G}_m$, GO_n, O_n, etc. for simplicity of notation.

When $n = 2m + 1$ is odd, the homomorphism $\mathbb{G}_m \times SO_{2m+1} \to GO_{2m+1}$ defined in terms of valued points by $(\lambda, g) \mapsto \lambda g$ is an isomorphism. This induces a 1-isomorphism of algebraic stacks $B\mathbb{G}_m \times BSO_{2m+1} \to BGO_{2m+1}$.

If X is a scheme over S, then a principal GO_n-bundle P on X, locally trivial in the étale topology, is equivalent to a nondegenerate quadratic triple (E, L, b). To any such P, we functorially associate the vector bundle E given by the defining representation $GO_n \hookrightarrow GL_n$, and L given by the character $\sigma : GO_n \to \mathbb{G}_m$ defined above. The standard bilinear form $\sum x_i y_i$ defines $b : E \otimes E \to L$. This defines a 1-isomorphism of stacks from the algebraic stack BGO_n to the algebraic stack $\mathcal{M} - \mathcal{D}$ of nondegenerate quadratic triples that is introduced later. The inverse 1-morphism has an obvious construction using the Gram-Schmidt orthogonalization process. The principal bundle P associated to any (E, L, b) is étale locally trivial as the process of Gram-Schmidt orthogonalization requires taking square-roots of invertible regular functions, which is possible étale locally as 2 is invertible over X.

Similarly, the stack BO_n can be identified with the stack of **nondegenerate**

quadratic pairs, where such a pair (E, q) on an S-scheme X is just a nondegenerate triple $(E, \mathcal{O}_X, q)$.

On the algebraic stack BGO_n, we have a universal nondegenerate quadratic triple (E_n, L_n, b_n). Let L_n also denote the total space of the line bundle L_n, so that L_n is again an algebraic stack with a projection $\pi : L_n \to BGO_n$ and a zero section $BGO_n \to L_n$, which are 1-morphisms of algebraic stacks. The stack L_n can be obtained from BGO_n by a local construction in the sense of Laumon, Moret-Bailly [L-MB] Chapter 14.

Let $L_n^\times \subset L_n$ be the open substack which is the complement of the zero section $BGO_n \hookrightarrow L_n$. For any X over S, an object of $L_n^\times(X)$ is a tuple (E, L, b, s) where (E, L, b) is a nondegenerate quadratic triple on X, and $s \in \Gamma(X, L)$ is a nowhere vanishing section. Equivalently, $s : \mathcal{O}_X \to L$ is an isomorphism. Therefore there is a 1-morphism of stacks $BO_n \to L_n^\times$, which associates to a nondegenerate quadratic pair $(F, q : F \otimes F \to \mathcal{O}_X)$ on X the tuple $(F, \mathcal{O}_X, q, \mathrm{id}_{\mathcal{O}_X})$. In the reverse direction, we can associate to a tuple $(E, L, b, s : \mathcal{O}_X \to L)$ the quadratic pair $(E, s^{-1} \circ b)$, showing that the 1-morphism $BO_n \to L_n^\times$ is an isomorphism of algebraic stacks. Using this, we identify $L_n^\times$ with BO_n.

As GO_n is a smooth group scheme over S, the stack BGO_n is smooth over S, hence the total space L_n is also smooth over S. We regard $BGO_n \hookrightarrow L_n$ as the zero section, so that (L_n, BGO_n) is a smooth pair over S, in particular, it is a regular pair. The projection $\pi : L_n \to BGO_n$ induces an isomorphism $H^*(BGO_n, \mathbb{F}_2) \to H^*(L_n, \mathbb{F}_2)$. Identifying $L_n^\times$ with BO_n, the Gysin sequence for (L_n, BGO_n) becomes the following exact sequence.

$$\ldots \to H^*(BGO_n, \mathbb{F}_2) \to H^*(BO_n, \mathbb{F}_2) \xrightarrow{d_n} H^{*-1}(BGO_n, \mathbb{F}_2) \to H^{*+1}(BGO_n, \mathbb{F}_2) \to \ldots$$

Characteristic classes for nondegenerate triples and $H^*(BGO_n)$

Given a group-scheme G over a base S, let $|BG|$ denote the set-valued functor on $(QuasiProj_S)^{opp}$ which associates to any quasi-projective S-scheme X the set of all isomorphism classes of principal G-bundles on X. Assuming that 2 is invertible on S, a **functorial characteristic class** for G will mean a natural transformation from $|BG|$ to the functor $H^*(-, \mathbb{F}_2)$ which associates to any X its étale cohomology $H^*(X, \mathbb{F}_2)$. All such functorial characteristic classes form a graded ring $CharClass_G^*$ under the ring operations on the various $H^*(X, \mathbb{F}_2)$.

If G is of finite-type and flat over S, then BG is an algebraic stack over S. Let $H^*(BG, \mathbb{F}_2)$ be its cohomology in the smooth-étale topology. We have a natural homomorphism

$$H^*(BG, \mathbb{F}_2) \to CharClass_G^*$$

of graded rings, which associates to any $\alpha \in H^*(BG, \mathbb{F}_2)$ the functorial characteristic class α' defined by $\alpha'(P) = \chi_P^*(\alpha)$ where $\chi_P : X \to BG$ is the classifying morphism of a G-bundle P on X.

When the base S is of the form $\mathrm{Spec}\, k$ for an algebraically closed field of characteristic $\neq 2$ and when G is reductive over k, it was shown in [Bh] that the homomorphism $H^*(BG, \mathbb{F}_2) \to CharClass_G^*$ is an isomorphism. We now revisit the argument in [Bh], to show it can be made to work over any excellent Dedekind domain on which 2 is invertible (for example, $\mathbb{Z}[1/2]$). This gives us the following.

Proposition 3.4.1. *Let G be a reductive group scheme over a base which is a field or an excellent Dedekind domain over $\mathbb{Z}[1/2]$. Then the natural homomorphism $H^*(BG, \mathbb{F}_2) \to CharClass_G^*$ is an isomorphism of graded rings.*

Proof (Sketch) The following three comments (1), (2) and (3) explain how the proof of the corresponding result (Proposition 3.1) in [Bh] can be modified so that it applies in our case.
(1) The Lemma 3.1 of [Bh] says that any closed reduced subscheme $Z \subset \mathbf{A}_k^n$ has a finite filtration by locally closed reduced subschemes $Z = Z_0 \supset Z_1 \supset \ldots \supset Z_\ell = \emptyset$ such that each $Z_i - Z_{i+1}$ is smooth over k. A modified statement holds for closed reduced subschemes $Z \subset \mathbf{A}_S^n$, with the condition '$Z_i - Z_{i+1}$ is smooth over k' replaced by the condition that $Z_i - Z_{i+1}$ is a regular scheme (we do not require $Z_i - Z_{i+1}$ to be smooth over S). Such a filtration exists because in any excellent scheme, regular points form an open subscheme (see Grothendieck [EGA-IV$_2$] page 215 statement 7.8.3.(iv)). An application of the absolute purity theorem of Gabber now completes the argument.
(2) The use of GIT for quotients for linear actions of reductive groups (in which which [Bh] follows Totaro [T]) is replaced by the use of Seshadri's version of GIT in [Se], which in particular works over the base S.
(3) Jouanolou's trick (in which [Bh] again follows Totaro [T]) remains valid over the base S (see [Ju]).

With the above modifications, the rest of the argument now works as in [Bh].
□

Cohomological invariants for nondegenerate quadric bundles

Tensor product of triples and line bundles defines a 1-morphism of algebraic stacks

$$\mu : BGO_n \times B\mathbb{G}_m \to BGO_n$$

which sends $(T,K) \mapsto T \otimes K$ where T is a quadratic triple on X and K is a line bundle on X. Following Toda [Td] as in [H-N-2], we say that an element $\alpha \in H^*(BGO_n, \mathbb{F}_2)$ is a **primitive class** if under the cohomology homomorphisms induced by μ and by the projection $p_1 : BGO_n \times B\mathbb{G}_m \to BGO_n$, we have

$$\mu^*(\alpha) = p_1^*(\alpha).$$

The primitive classes form a subring $PH^*(BGO_n, \mathbb{F}_2) \subset H^*(BGO_n, \mathbb{F}_2)$. Over a field or an excellent Dedekind domain S over $\mathbb{Z}[1/2]$, the arguments used for proving that the natural homomorphism $H^*(BG, \mathbb{F}_2) \to CharClass_G^*$ is an isomorphism can be used to see that elements $\alpha \in PH^*(BGO_n, \mathbb{F}_2)$ can also be characterized as natural transformations from $|BGO_n|$ to $H^*(-, \mathbb{F}_2)$ on $QuasiProj_S$ such that $\alpha(T \otimes K) = \alpha(T)$ for any nondegenerate triple T and a line bundle K on a quasi-projective scheme X over S.

3.5 The main theorem

3.5.1 Statement of the main theorem

We are now able to state the main theorem, having made all the above preparation over étale cohomology and Gysin sequences for algebraic stacks. But the statement is just the algebraic stacky analogue of the corresponding topological theorem in [H-N-2], where the complex Lie groups $GO_n(\mathbb{C})$, etc. are replaced by the corresponding group schemes over S, and singular cohomology with coefficients $\mathbb{F}_2$ is replaced by étale cohomology with coefficients $\mathbb{F}_2$. We will adhere to the notation in [H-N-2] as much as possible.

Let $PH^*(BGO_n, \mathbb{F}_2) \subset H^*(BGO_n, \mathbb{F}_2)$ be the subring of primitive classes. Its elements are the universal cohomological invariants for nondegenerate quadric bundles of rank n. Let $B(v)^* : H^*(BGO_n) \to H^*(BO_{n-1})$ be the ring homomorphism induced by the group homomorphism $v : O_{n-1} \to GO_n$ defined by $g \mapsto \begin{pmatrix} 1 & \\ & g \end{pmatrix}$. Let $(E_{n-1}, L_{n-1}, b_{n-1})$ be the universal triple over BGO_{n-1}. As shown above, the complement $L_{n-1}^{\times}$ of the zero section of L_{n-1} is isomorphic to BO_{n-1}, and there is a Gysin boundary map $d_{n-1} : H^*(BO_{n-1}) \to H^{*-1}(BGO_{n-1})$. Finally, let $\delta : PH^*(BGO_n) \to H^{*-1}(BGO_{n-1})$ be the composite linear map

$$PH^*(BGO_n) \hookrightarrow H^*(BGO_n) \stackrel{B(v)^*}{\to} H^*(BO_{n-1}) \stackrel{d_{n-1}}{\to} H^{*-1}(BGO_{n-1})$$

With the above notations, we have the following, where all cohomologies are smooth étale cohomologies with coefficients $\mathbb{F}_2$.

Theorem 3.5.1. *(***Main Theorem***) Let S be a regular scheme over $\mathbb{Z}[1/2]$. Let (X,Y) be a regular pair of stacks over S. Let Q be a quadric bundle on X, nondegenerate on $X-Y$ of rank $n \geq 2$, which is minimally degenerate on $Y \subset X$. Let $\mathbb{T}^{Q_Y}$ be the associated rank $n-1$ nondegenerate quadratic triple on Y. Let $\alpha \in PH^*(BGO_n)$ be a universal quadric invariant, and let $\alpha(Q_{X-Y}) \in H^*(X-Y)$ be its value on Q_{X-Y}. Let $\delta : PH^*(BGO_n) \to H^{*-1}(BGO_{n-1})$ be the linear map defined above, and let $(\delta(\alpha))(\mathbb{T}^{Q_Y})$ be the value of the resulting GO_{n-1}-characteristic class $\delta(\alpha)$ on $\mathbb{T}^{Q_Y}$. Let $\nu_Y(\det(Q)) \in H^0(Y)$ be the vanishing multiplicity along Y of the discriminant $\det(Q) \in \Gamma(X, L^n \otimes \det(E)^{-2})$. Then under the Gysin boundary map $\partial : H^*(X-Y) \to H^{*-1}(Y)$, we have the equality*

$$\partial(\alpha(Q_{X-Y})) = \nu_Y(\det(Q)) \cdot (\delta(\alpha))(\mathbb{T}^{Q_Y})$$

The rest of this paper is devoted to proving the theorem.

3.5.2 The algebraic stack of quadratic triples

Let $n \geq 1$. For any S-scheme X, let $\mathcal{M}_n(X)$ be the groupoid whose objects are all quadratic triples $T = (E, L, b)$ on X with $\text{rank}(E) = n$, and a morphism $T \to T'$, where $T = (E, L, b)$ and $T' = (E', L', b')$, is pair of $\mathcal{O}_X$-linear isomorphisms $E \to E'$ and $L \to L'$ which takes b to b'. If $f : X' \to X$ is a morphism of S-schemes and T is a quadratic triple over X, the pull-back triple $f^*(T) = (f^*E, f^*L, f^*b)$ is defined in the obvious manner. For each $f : X' \to X$ and $f'' : X'' \to X'$ over S, we can define an isomorphism $\phi_{f,f'} : (ff')^*(T) \to f'^* f^*(T)$, which makes $\mathcal{M}_n$ into a groupoid-valued pseudo functor on S-schemes. It is clear that the pull-back f^* preserves both the dimension (rank of E) and the point-wise ranks (ranks of b_x), and pulls back the discriminant, so properties such as nondegenerateness, minimal degenerateness, etc. are preserved under pullbacks. The **almost nondegenerate triples**, which will mean triples whose point-wise rank is $\geq n-1$, form a full subgroupoid $\mathcal{M}$ of $\mathcal{M}_n$. The S-groupoid $\mathcal{M}$ has a full subgroupoid $\mathcal{D}$ formed by all the quadratic triples $T = (E, L, b)$ on S-schemes X such that $\det(T) = 0 \in \Gamma(X, L^n \otimes \det(E)^{-2})$ and $\text{rank}(b_x) = n-1$ for all $x \in X$. This is the groupoid of minimally degenerate triples.

Proposition 3.5.2. *For any $n \geq 1$, almost nondegenerate triples of dimension n form an algebraic stack $\mathcal{M}$ over S, and minimally degenerate triples of dimension n form a closed substack $\mathcal{D} \subset \mathcal{M}$ defined by the vanishing of the discriminant of the universal triple $\mathcal{T}$ on $\mathcal{M}$. Moreover, $\mathcal{D}$ is a smooth relative divisor in $\mathcal{M}$ over S. The stack $\mathcal{M}$ is a global quotient stack under the reductive group scheme $GL_n \times \mathbb{G}_m$ over S.*

Proof We will see below that the S-groupoid $\mathcal{M}_n$ of all triples is an algebraic stack on S. It has a filtration by closed algebraic substacks $\mathcal{M}_n = \mathcal{M}_n^{\leq n} \supset \mathcal{M}_n^{\leq n-1} \supset \mathcal{M}_n^{\leq n-2} \supset \ldots$ where for any X, $\mathcal{M}_n^{\leq r}(X)$ is the groupoid of all triples (E, L, b) on X such that the determinants of minors of b of size $r+1$ vanish identically on X (the relevant Fitting ideal is 0). The almost nondegenerate triples form the open substack

$$\mathcal{M} = \mathcal{M}_n - \mathcal{M}_n^{\leq n-2}.$$

The closed substack $\mathcal{D} \subset \mathcal{M}$ is given by

$$\mathcal{D} = \mathcal{M}_n^{\leq n-1} - \mathcal{M}_n^{\leq n-2}.$$

We now construct these stacks $\mathcal{M}_n^{\leq r}$. Let E be the universal vector bundle on the stack BGL_n and let L be the universal line bundle on $B\mathbb{G}_m$. On the stack $BGL_n \times B\mathbb{G}_m$ we get the vector bundles $p_1^*\underline{Sym}^2(E)$ and p_2^*L. Then $\mathcal{M}_n$ is the 'total space' (rather,'total stack') of the vector bundle corresponding to the locally free $\mathcal{O}$-module $\underline{Hom}(p_1^*\underline{Sym}^2(E), p_2^*L)$ on $BGL_n \times B\mathbb{G}_m$, which is an algebraic stack as it is a local construction in the sense of Laumon, Moret-Bailly [L-MB] Chapter 14. The closed substack $\mathcal{M}_n^{\leq r}(X)$ is defined by the vanishing of all determinants of minors of size $r+1$. In particular, $\mathcal{D}$ is defined in $\mathcal{M}$ by the vanishing of the discriminant, so it is a principal divisor.

As $BGL_n \times B\mathbb{G}_m$ is smooth over S, and as $\mathcal{M}_n$ is a geometric vector bundle over it, $\mathcal{M}_n$ is a smooth stack over S. The Jacobian criterion applied to the discriminant shows that $\mathcal{D}$ is a smooth relative divisor in $\mathcal{M}$ over S.

The affine space $\mathbf{A}_S^{n(n+1)/2}$ over S of $n \times n$-symmetric matrices has an action of $GL_n \times \mathbb{G}_m$ given in terms of valued points by $b \cdot (g, \lambda) = {}^t g b g \lambda^{-1}$. If $M \subset \mathbf{A}_S^{n(n+1)/2}$ is the open subscheme where $\mathrm{rank}(b) \geq n-1$, and if $D \subset M$ is the divisor defined by $\det(b) = 0$, then it can be seen that $\mathcal{M}$ is isomorphic to the quotient stack $[M/GL_n \times \mathbb{G}_m]$ under which $\mathcal{D}$ becomes its the closed substack $[D/GL_n \times \mathbb{G}_m]$. □

Note that we have a tautological quadratic triple $\mathcal{T}_n = (\mathcal{E}_n, \mathcal{L}_n, \beta_n)$ on $(\mathcal{M}, \mathcal{D})$. If X is an algebraic stack, then to any almost nondegenerate quadratic triple T on X there functorially corresponds its **classifying morphism** $\chi_T : X \to \mathcal{M}$, together with an isomorphism of triples $\chi_T^*(\mathcal{T}_n) \to T$. The degeneration multiplicity along Y of a mildly degenerating triple T on a regular pair (X, Y) of stacks has the following obvious interpretation in terms of its classifying morphism $\chi_T : X \to \mathcal{M}$.

Proposition 3.5.3. *Let $T = (E, L, b)$ be a mildly degenerating triple on a regular pair (X, Y), with degeneration multiplicity $\nu_Y(T) : |Y| \to \mathbb{Z}_{\geq 0}$. Then under the* *classifying morphism $\chi_T : X \to \mathcal{M}$, the pull back of the closed substack $\mathcal{D} \subset \mathcal{M}$ is the closed substack $\nu_Y(T)Y \subset X$ defined by the ideal sheaf $(I_Y)^{\nu_Y(T)} \subset \mathcal{O}_X$.* □

We now translate the main theorem in terms of the Gysin homomorphism $\partial : H^*(\mathcal{M} - \mathcal{D}) \to H^{*-1}(\mathcal{D})$. For this, we first define a 1-morphism $\rho : \mathcal{D} \to BGO_{n-1}$ as follows. If Y is any (affine) scheme over S, an object of the groupoid $\mathcal{D}(Y)$ is a minimally degenerate triple $T = (F, L, q)$ over Y, with $\mathrm{rank}(F) = n$ and $rank(q) = n - 1$. Note that $\ker(q) \subset F$ is a line subbundle, and q induces a nondegenerate L-valued form $\overline{q}$ on the quotient $F/\ker(q)$. We get a new nondegenerate triple of rank $n - 1$

$$\rho(F, L, q) = (F/\ker(q), L, \overline{q}) \otimes \ker(q)^{-1}$$

over Y. This is functorial in the triples and respects their pull-backs, so it defines 1-morphism of stacks $\rho : \mathcal{D} \to BGO_{n-1}$. Note that if $Q = [T]$ is a mildly degenerating quadric bundle on (X, Y), then

$$\mathbb{T}^{Q_Y} = \rho(T|_Y).$$

Hence the 1-morphism $\rho : \mathcal{D} \to BGO_{n-1}$ is the classifying morphism of the triple $\mathbb{T}^{\mathcal{Q}_\mathcal{D}}$ on $\mathcal{D}$, where $\mathcal{Q}$ is the mildly degenerating quadric bundle $\mathcal{Q} = [\mathcal{T}]$ on $(\mathcal{M}, \mathcal{D})$ corresponding to the universal triple $\mathcal{T}$ on $\mathcal{M}$.

We will prove the main theorem in the following equivalent re-formulation.

Theorem 3.5.4. Main theorem restated in terms of the pair $(\mathcal{M}, \mathcal{D})$: *The Gysin boundary map $\partial : H^*(\mathcal{M}-\mathcal{D}) \to H^{*-1}(\mathcal{D})$ on the primitive cohomology $PH^*(\mathcal{M} - \mathcal{D}) \subset H^*(\mathcal{M} - \mathcal{D})$ is given by the formula*

$$\partial|_{PH^*(\mathcal{M}-\mathcal{D})} = \rho^* \circ d_{n-1} \circ B(v)^*$$

where $B(v)^* : H^*(\mathcal{M}-\mathcal{D}) = H^*(BGO_n) \to H^*(BO_{n-1})$ is induced by $v : O_{n-1} \to GO_n : g \mapsto \begin{pmatrix} 1 & \\ & g \end{pmatrix}$, the map $d_{n-1} : H^*(BO_{n-1}) \to H^{*-1}(BGO_{n-1})$ is the Gysin boundary map for the pair (BO_{n-1}, BGO_{n-1}), and $\rho : \mathcal{D} \to BGO_{n-1}$ is the 1-morphism defined by $\rho(F, L, q) = (F/\ker(q), L, \overline{q}) \otimes \ker(q)^{-1}$.

Proof of equivalence: Consider the universal quadratic triple $\mathcal{T}$ on $\mathcal{M}$, and the resulting quadric bundle $\mathcal{Q} = [\mathcal{T}]$, which is a mildly degenerating quadric bundle on the regular pair of stacks $(\mathcal{M}, \mathcal{D})$, with degeneration multiplicity $\nu_\mathcal{D}(\det(\mathcal{Q})) = 1$. Let $\alpha \in PH^*(\mathcal{M} - \mathcal{D}) = PH^*(BGO_n)$. By definition, $\alpha(\mathcal{Q}_{\mathcal{M}-\mathcal{D}}) = \alpha$. Therefore the formula $\partial(\alpha) = \nu_Y(\det(Q)) \cdot (\delta(\alpha))(\mathbb{T}^{Q_Y})$ of the Theorem 3.5.1 applied to α and $\mathcal{Q}$ becomes $\partial(\alpha(\mathcal{Q}_{\mathcal{M}-\mathcal{D}})) = \delta(\alpha)(\mathbb{T}^{\mathcal{Q}_\mathcal{D}})$. As remarked above,

$\mathbb{T}^{Q_{\mathcal{D}}} = \rho(\mathcal{T}|_{\mathcal{D}})$. Hence we get for all $\alpha \in PH^*(\mathcal{M} - \mathcal{D})$ the equality $\partial(\alpha) = \rho^*\delta(\alpha) = \rho^* \circ d_{n-1} \circ B(v)^*(\alpha)$ by definition of the map δ from Theorem 3.5.1. As this holds for all $\alpha \in PH^*(\mathcal{M} - \mathcal{D})$, we get $\partial|_{PH^*(\mathcal{M}-\mathcal{D})} = \rho^* \circ d_{n-1} \circ B(v)^*$ as desired. (The above is a typical 'Yoneda argument'.)

Conversely, suppose the above formula holds. Given any mildly degenerating quadric Q on a regular pair of stacks (X, Y), let $\chi : X \to \mathcal{M}$ be the classifying 1-morphism for any chosen quadratic triple T on X which represents Q. Let $m = \nu_Y(Q)$. By Lemma 3.2.5.(2) we get a commutative diagram

$$\begin{array}{ccc} H^i(\mathcal{M} - \mathcal{D}) & \stackrel{\partial_{(\mathcal{M},\mathcal{D})}}{\to} & H^{i-1}(\mathcal{D}) \\ \chi|^*_{X-Y} \downarrow & & \downarrow m\cdot\chi|^*_Y \\ H^i(X - Y) & \stackrel{\partial_{(X,Y)}}{\to} & H^{i-1}(Y) \end{array}$$

We now make the following observation (3.5.5) before continuing with the proof that (3.5.1) implies (3.5.4).

3.5.5 Note that for any $\alpha \in H^i(\mathcal{M} - \mathcal{D})$, by definition $\alpha(T) = \chi|^*_{X-Y}(\alpha)$, and in particular when $\alpha \in PH^i(\mathcal{M} - \mathcal{D})$, by definition $\alpha(Q) = \chi|^*_{X-Y}(\alpha)$. Also, for any $\beta \in H^{i-1}(\mathcal{D})$, by definition $\beta(T|_Y) = \chi|^*_Y(\beta)$.

For any $\alpha \in PH^*(\mathcal{M} - \mathcal{D})$ we therefore have the following sequence of equalities.

$$\begin{aligned} \partial_{(X,Y)}(\alpha(Q)) &= \partial_{(X,Y)}\chi|^*_{X-Y}(\alpha) \text{ by the Statement 3.5.5,} \\ &= m \cdot \chi|^*_Y\partial_{(\mathcal{M},\mathcal{D})}(\alpha) \text{ by the above application of Lemma 3.2.5.(2),} \\ &= m \cdot \chi|^*_Y\partial|_{PH^*(\mathcal{M}-\mathcal{D})}(\alpha) \text{ as by assumption } \alpha \in PH^*(\mathcal{M} - \mathcal{D}), \\ &= m \cdot \chi|^*_Y\rho^* \circ d_{n-1} \circ B(v)^*(\alpha) \text{ if Theorem 3.5.4 holds,} \\ &= m \cdot \chi|^*_Y\rho^*\delta(\alpha), \\ &\quad \text{where } \delta : PH^*(\mathcal{M} - \mathcal{D}) = PH^*(BGO_n) \to H^{*-1}(BGO_{n-1}) \\ &\quad \text{is defined as earlier by } \delta = d_{n-1} \circ B(v)^*. \end{aligned}$$

This gives the formula

$$\partial_{(X,Y)}(\alpha(Q)) = m \cdot \chi|^*_Y\rho^*\delta(\alpha)$$

Note that the composite 1-morphism $Y \stackrel{\chi|_Y}{\to} \mathcal{D} \stackrel{\rho}{\to} BGO_{n-1}$ is exactly the characteristic morphism $\chi' : Y \to BGO_{n-1}$ of the reduced triple $\mathbb{T}^{Q_Y} = (E/\ker(b), L, \bar{b}) \otimes \ker(b)^{-1}$ where $T = (E, L, b)$ represents Q. Hence we get

$$\chi|^*_Y\rho^*\delta(\alpha) = (\chi')^*\delta(\alpha) = \delta(\alpha)(\mathbb{T}^{Q_Y})$$

Putting together the previous two displayed equalities, we get

$$\partial_{(X,Y)}(\alpha(Q)) = \delta(\alpha)(\mathbb{T}^{Q_Y})$$

as desired. This completes the proof of the equivalence of Theorems 3.5.1 and 3.5.4. $\square$

3.5.3 Standard models for degenerating quadrics

Let K' be the universal line bundle on $B\mathbb{G}_m$. Consider any $n \geq 1$ and the universal triple $(E_{n-1}, L_{n-1}, b_{n-1})$ on the stack BGO_{n-1}. Consider the projections $p_1 : BGO_{n-1} \times B\mathbb{G}_m \to BGO_{n-1}$ and $p_2 : BGO_{n-1} \times B\mathbb{G}_m \to B\mathbb{G}_m$. Let $(\mathbf{E}, \mathbf{L}, \mathbf{b}) = p_1^*(E_{n-1}, L_{n-1}, b_{n-1})$ and $K = p_2^* K'$ be the pullbacks. Let the stack $\mathcal{N}$ be the total space of the line bundle $\mathbf{L}$ on the stack $\mathcal{Z} = BGO_{n-1} \times B\mathbb{G}_m$, and let $\pi : \mathcal{N} \to \mathcal{Z}$ be the projection 1-morphism. We regard $\mathcal{Z}$ as a closed substack $\mathcal{Z} \hookrightarrow \mathcal{N}$, embedded by the zero section. Note that the stacks $\mathcal{Z}$ and $\mathcal{N}$ are smooth over S.

We now define a certain quadratic triple on $\mathcal{N}$ following [H-N-2], called the **model triple**. Let $\tau \in \Gamma(\mathcal{N}, \pi^*(\mathbf{L}))$ be the tautological section, which vanishes exactly on the zero section $BGO_{n-1} \times B\mathbb{G}_m = \mathcal{Z} \subset \mathcal{N}$, with vanishing multiplicity 1. Then we get a quadratic triple (where $\oplus$ denotes the orthogonal direct sum)

$$T = ((\mathcal{O}_{\mathcal{N}}, \pi^*\mathbf{L}, \tau) \oplus \pi^*(\mathbf{E}, \mathbf{L}, \mathbf{b})) \otimes K$$

on $\mathcal{N}$ of dimension n, which has rank n on $\mathcal{N} - \mathcal{Z}$ and has rank $n-1$ on $\mathcal{Z}$, which means it is a mildly degenerating triple on $(\mathcal{N}, \mathcal{Z})$, which is nondegenerate on $\mathcal{N} - \mathcal{Z}$ and minimally degenerate on the divisor $\mathcal{Z}$, with degeneration multiplicity $\nu_{\mathcal{Z}}(\det(T)) = 1$.

Let $\varphi : \mathcal{Z} \to \mathcal{D}$ be the classifying morphism of the degenerate triple

$$T_0 = ((\mathcal{O}_{\mathcal{Z}}, \mathbf{L}, 0) \oplus (\mathbf{E}, \mathbf{L}, \mathbf{b})) \otimes K$$

which is the restriction of T to $\mathcal{Z} \hookrightarrow \mathcal{N}$. In the opposite direction, let the 1-morphism $\psi : \mathcal{D} \to \mathcal{Z}$ be defined as follows. For any S-scheme Y, an object of $\mathcal{D}(Y)$ is a minimally degenerate quadratic triple (F, L, q) on Y. Hence $\ker(q)$ is a line bundle on Y so it is an object of $B\mathbb{G}_m(Y)$, while $(F/\ker(q), L, \overline{q})$ (where $\overline{q}$ is induced by q) is an object of $BGO_{n-1}(Y)$. As $\mathcal{Z} = BGO_{n-1} \times B\mathbb{G}_m$, we can define

$$\psi(F, L, q) = ((F/\ker(q), L, \overline{q}) \otimes \ker(q)^{-1}, \ker(q)) \in Ob\, \mathcal{Z}(Y).$$

3.5.6 (Factorization of ρ) The 1-morphism $\rho : \mathcal{D} \to BGO_{n-1}$, defined earlier by $\rho(F, L, q) = (F/\ker(q), L, \overline{q}) \otimes \ker(q)^{-1}$, factors as

$$\rho = p_1 \circ \psi$$

where $p_1 : BGO_{n-1} \times B\mathbb{G}_m \to BGO_{n-1}$ is the projection.

The composite 1-morphism $\varphi \circ \psi : \mathcal{D} \to \mathcal{D}$ is of importance to us. In terms of the above notation, we have

$$\varphi \circ \psi(F, L, q) = (\ker(q), L, 0) \oplus (F/\ker(q), L, \overline{q}).$$

Lemma 3.5.7. *The 1-morphisms $\varphi : \mathcal{Z} \to \mathcal{D}$ and $\psi : \mathcal{D} \to \mathcal{Z}$ satisfy the following.*
(1) $\psi \circ \varphi = \mathrm{id}_{\mathcal{Z}}$.
(2) $\varphi \circ \psi : \mathcal{D} \to \mathcal{D}$ is $\mathbf{A}^1$-homotopic to $\mathrm{id}_{\mathcal{D}}$, that is, there exists a 1-morphism $F : \mathcal{D} \times \mathbf{A}^1 \to \mathcal{D}$ such that $\varphi \circ \psi = F_0$ and $\mathrm{id}_{\mathcal{D}} = F_1$, where F_t denotes $F|_{\mathcal{D} \times \{t\}}$ for $t \in \mathbf{A}^1(\mathbb{Z}[1/2]) = \mathbb{Z}[1/2]$. Hence by Remark 3.2.7, the cohomology map $\psi^* \circ \varphi^* : H^*(\mathcal{D}) \to H^*(\mathcal{D})$ is identity.
(3) Consequently, the cohomology maps $\psi^* : H^*(\mathcal{Z}) \to H^*(\mathcal{D})$ and $\varphi^* : H^*(\mathcal{D}) \to H^*(\mathcal{Z})$ are isomorphisms, which are inverse to each other.

Proof Let $0 \to E' \to E \to E'' \to 0$ is a short exact sequence of vector bundles on an S-scheme Y. Let $\mathbf{A}^1$ be the affine line over with coordinate t. On $Y \times \mathbf{A}^1$ consider the homomorphism $g : p_Y^* E' \to p_Y^* E' \oplus p_Y^* E$ which sends $v \mapsto (tv, v)$. Then $\mathcal{F} = \mathrm{coker}(g)$ fits in a short exact sequence $0 \to p_Y^* E' \to \mathcal{F} \to p_Y^* E'' \to 0$ of vector bundles on $Y \times \mathbf{A}^1$, which specializes to the given exact sequence on Y for $t = 1$ and specializes to the split exact sequence $0 \to E' \to E' \oplus E'' \to E'' \to 0$ for $t = 0$. The above construction is functorial, hence it also works for short exact sequences of vector bundles over algebraic stacks.

Applying this to the short exact sequence $0 \to \ker(b_n|_{\mathcal{D}}) \to \mathcal{E}_n|_{\mathcal{D}} \to \overline{\mathcal{E}_n|_{\mathcal{D}}} \to 0$ on $\mathcal{D}$ which arises from the universal minimally degenerate triple $\mathcal{T}_{\mathcal{D}} = (\mathcal{E}_n, \mathcal{L}_n, \beta_n)|_{\mathcal{D}}$, we get a short exact sequence $0 \to p_{\mathcal{D}}^* \ker(b_n|_{\mathcal{D}}) \to \mathcal{F} \to p_{\mathcal{D}}^* \overline{\mathcal{E}_n|_{\mathcal{D}}} \to 0$ of vector bundles on $\mathcal{D} \times \mathbf{A}^1$. Let $\widehat{b} : \mathcal{F} \otimes \mathcal{F} \to p_{\mathcal{D}}^* L$ be induced by $p_{\mathcal{D}}^* \overline{b}$. This defines a minimally degenerate triple $\widehat{\mathcal{T}_{\mathcal{D}}} = (\mathcal{F}, p_{\mathcal{D}}^* L, \widehat{b})$ on $\mathcal{D} \times \mathbf{A}^1$. Take $F = \chi_{\widehat{\mathcal{T}_{\mathcal{D}}}} : \mathcal{D} \times \mathbf{A}^1 \to \mathcal{D}$ to be its classifying morphism. □

Let $\chi : \mathcal{N} \to \mathcal{M}$ be the classifying morphism of the mildly degenerating quadratic triple $T = ((\mathcal{O}_{\mathcal{N}}, \pi^* \mathbf{L}, \tau) \oplus \pi^*(\mathbf{E}, \mathbf{L}, \mathbf{b})) \otimes K$ on $(\mathcal{N}, \mathcal{Z})$ defined above. Under it, the pullback of $\mathcal{D} \subset \mathcal{M}$ is the zero section $\mathcal{Z} \subset \mathcal{N}$, with multiplicity 1. Both the pairs $(\mathcal{M}, \mathcal{D})$ and $(\mathcal{N}, \mathcal{Z})$ admit long exact Gysin sequences, and we have a commutative diagram as follows, where the vertical maps are induced by the classifying morphism of T and its restrictions.

$$\begin{array}{ccccccccc} \ldots \to & H^i(\mathcal{M}) & \to & H^i(\mathcal{M}-\mathcal{D}) & \stackrel{\partial}{\to} & H^{i-1}(\mathcal{D}) & \to & H^{i+1}(\mathcal{M}) & \to \ldots \\ & \downarrow \chi^* & & \downarrow \chi|^*_{\mathcal{N}-\mathcal{Z}} & & \downarrow \chi|^*_{\mathcal{Z}} & & \downarrow \chi^* & \\ \ldots \to & H^i(\mathcal{N}) & \to & H^i(\mathcal{N}-\mathcal{Z}) & \stackrel{d}{\to} & H^{i-1}(\mathcal{Z}) & \to & H^{i+1}(\mathcal{N}) & \to \ldots \end{array}$$

Note that $\mathcal{M}-\mathcal{D}$ is isomorphic to BGO_n, and $\mathcal{N}-\mathcal{Z}$ is isomorphic to $BO_{n-1} \times B\mathbb{G}_m$. Under these isomorphisms, the above map $\chi|^*_{\mathcal{N}-\mathcal{Z}} : H^i(\mathcal{M}-\mathcal{D}) \to H^i(\mathcal{N}-\mathcal{Z})$ is the cohomology map $H^i(BGO_n) \to H^i(BO_{n-1} \times B\mathbb{G}_m)$ associated to the 1-morphism $BO_{n-1} \times B\mathbb{G}_m \to BGO_n$ of stacks which is induced by the group scheme homomorphism

$$V : O_{n-1} \times \mathbb{G}_m \to GO_n : (g, \lambda) \mapsto \begin{pmatrix} 1 & \\ & g \end{pmatrix} \lambda.$$

Hence the commutative diagram

$$\begin{array}{ccc} H^i(\mathcal{M}-\mathcal{D}) & \stackrel{\partial_{(\mathcal{M},\mathcal{D})}}{\to} & H^{i-1}(\mathcal{D}) \\ \downarrow \chi|^*_{\mathcal{N}-\mathcal{Z}} & & \downarrow \chi|^*_{\mathcal{Z}} \\ H^i(\mathcal{N}-\mathcal{Z}) & \stackrel{\partial_{(\mathcal{N},\mathcal{Z})}}{\to} & H^{i-1}(\mathcal{Z}) \end{array}$$

becomes the commutative diagram

$$\begin{array}{ccc} H^i(BGO_n) & \stackrel{\partial_{(\mathcal{M},\mathcal{D})}}{\to} & H^{i-1}(\mathcal{D}) \\ B(V)^* \downarrow & & \downarrow \varphi^* \\ H^i(BO_{n-1} \times B\mathbb{G}_m) & \stackrel{d}{\to} & H^{i-1}(BGO_{n-1} \times B\mathbb{G}_m) \end{array}$$

3.5.8 (Factorization of the Gysin map $\partial : H^i(\mathcal{M}-\mathcal{D}) \to H^{i-1}(\mathcal{D})$) As φ^* and ψ^* are inverses by Lemma 3.5.7, the above commutative diagram gives the following commutative diagram.

$$\begin{array}{ccc} H^i(BGO_n) & \stackrel{\partial_{(\mathcal{M},\mathcal{D})}}{\to} & H^{i-1}(\mathcal{D}) \\ B(V)^* \downarrow & & \uparrow \psi^* \\ H^i(BO_{n-1} \times B\mathbb{G}_m) & \stackrel{d}{\to} & H^{i-1}(BGO_{n-1} \times B\mathbb{G}_m) \end{array}$$

Lemma 3.5.9. *Let $PH^*(BGO_n) \subset H^*(BGO_n)$ be the primitive subring. Then we have a commutative diagram*

$$\begin{array}{ccc} PH^i(BGO_n) & \hookrightarrow & H^i(BGO_n) \\ \downarrow B(v)^* & & \downarrow B(V)^* \\ H^i(BO_{n-1}) & \stackrel{p_1'^*}{\to} & H^i(BO_{n-1} \times B\mathbb{G}_m) \end{array}$$

where $p_1'^*$ is induced by the projection $p_1' : BO_{n-1} \times B\mathbb{G}_m \to BO_{n-1}$.

Proof Let (F, q) be the universal quadratic form on BO_{n-1}, where F is a vector bundle of rank $n-1$ and q is a nondegenerate quadratic form on F with values in $\mathcal{O}_{BO_{n-1}}$. Let K be the universal line bundle on $B\mathbb{G}_m$. The morphism $B(V) : BO_{n-1} \times B\mathbb{G}_m \to BGO_n$ is the classifying map of the nondegenerate quadratic triple $T = ((\mathcal{O}, \mathcal{O}, 1) \oplus (F, \mathcal{O}, q)) \otimes K$. If $\alpha \in H^*(BGO_n)$ then $B(V)^*(\alpha) = \alpha(T)$. As $\alpha \in PH^*(BGO_n)$, we have $\alpha(T) = \alpha(T \otimes K^{-1}) = \alpha((\mathcal{O}, \mathcal{O}, 1) \oplus (F, \mathcal{O}, q)) = B(v)^*(\alpha)$. □

3.5.4 Completion of the proof of the main theorem

3.5.10 Let $p_1 : BGO_{n-1} \times B\mathbb{G}_m \to BGO_{n-1}$ and $p'_1 : BO_{n-1} \times B\mathbb{G}_m \to BO_{n-1}$ be the projections. Then the following diagram commutes, where the horizontal maps are Gysin boundary maps.

$$\begin{array}{ccc} H^i(BO_{n-1}) & \stackrel{d_{n-1}}{\to} & H^{i-1}(BGO_{n-1}) \\ p'^*_1 \downarrow & & \downarrow p_1^* \\ H^i(BO_{n-1} \times B\mathbb{G}_m) & \stackrel{d}{\to} & H^{i-1}(BGO_{n-1} \times B\mathbb{G}_m) \end{array}$$

As before, we identify $\mathcal{M} - \mathcal{D}$ with BGO_n, under which $PH^*(\mathcal{M} - \mathcal{D})$ gets identified with $PH^*(BGO_n) \subset H^*(BGO_n)$. We then have the following sequence of equalities.

$$\begin{aligned} \partial|_{PH^*(BGO_n)} &= \psi^* \circ d \circ B(V)^* \text{ by Statement 3.5.8,} \\ &= \psi^* \circ d \circ p'^*_1 \circ B(v)^* \text{ by Lemma 3.5.9,} \\ &= \psi^* \circ p_1^* \circ d_{n-1} \circ B(v)^* \text{ by Statement 3.5.10,} \\ &= \rho^* \circ d_{n-1} \circ B(v)^* \text{ by Statement 3.5.6.} \end{aligned}$$

Hence the Gysin boundary map $\partial_{(\mathcal{M},\mathcal{D})} : H^*(\mathcal{M} - \mathcal{D}) \to H^{*-1}(\mathcal{D})$ satisfies the formula

$$\partial|_{PH^*(\mathcal{M}-\mathcal{D})} = \rho^* \circ d_{n-1} \circ B(v)^*$$

which completes the proof of Theorem 3.5.4, and hence of Theorem 3.5.1. □

3.5.5 The case of a separably closed base field $k = k_{sep}$

When our base S is of the form Spec k for a separably closed field k of characteristic $\neq 2$, the rings $H^*(BGO_{n,k}, \mathbb{F}_2)$ and $PH^*(BGO_{n,k}, \mathbb{F}_2)$ are the same as in the topological case by [Bh], so the Theorem 3.5.1 now implies that all the calculations in [H-N-2] apply without change over such a k.

In [Bh], it was shown that the key topological facts on which the the calculation of $H^*_{sing}(BGO_{2n}(\mathbb{C}), \mathbb{F}_2)$ in [H-N-1] is based have suitable analogs in the algebraic category over a separably closed base field k of characteristic $\neq 2$, and hence we get the following determination of $H^*(BGO_{n,k}, \mathbb{F}_2)$.

First, recall that the cohomology ring $H^*(BGL_{n,k}, \mathbb{F}_2)$ is the polynomial ring $\mathbb{F}_2[\bar{c}_1, \ldots, \bar{c}_n]$ in n variables $\bar{c}_i \in H^{2i}(BGL_{n,k}, \mathbb{F}_2)$ which are the **universal mod 2 Chern classes**. Also, the cohomology ring $H^*(BO_{n,k}, \mathbb{F}_2)$ is the polynomial ring $\mathbb{F}_2[w_1, \ldots, w_n]$ in n variables $w_i \in H^i(BGL_{n,k}, \mathbb{F}_2)$ which are the **universal Stiefel-Whitney classes** of Laborde [La] over $k = k_{sep}$ (see, for example, [E-K-V]). Under the homomorphism $H^*(BGL_{n,k}, \mathbb{F}_2) \to H^*(BO_{n,k}, \mathbb{F}_2)$ induced by the inclusion $O_{n,k} \hookrightarrow GL_{n,k}$, we have $\bar{c}_i \mapsto w_i^2$.

3.5.11 The rings $H^*(BGO_{2n+1,k}, \mathbb{F}_2)$. For the odd special orthogonal group scheme $SO_{2n+1,k}$ we have $H^*(BSO_{2n+1,k}, \mathbb{F}_2) = \mathbb{F}_2[w_2, \ldots, w_{2n+1}]$, as follows from the direct product decomposition $O_{2n+1} = \mu_2 \times SO_{2n+1,k}$ by using the Künneth formula (see section 5 of [Bh] for details of the argument). We have an isomorphism $\mathbb{G}_m \times SO_{2n+1} \to GO_{2n+1}$ defined in terms of valued points by $(\lambda, g) \mapsto \lambda g$. It follows that $BGO_{2n+1} = B\mathbb{G}_m \times BSO_{2n+1}$. Hence we have

$$H^*(BGO_{2n+1,k}, \mathbb{F}_2) = \mathbb{F}_2[c, w_2, \ldots, w_{2n+1}]$$

where c denotes the pullback of the first Chern class $\bar{c}_1$ from $H^2(BG_{m,k}, \mathbb{F}_2)$, and the $w_2, \ldots, w_{2n+1}$ are the pull-backs of the Hasse-Witt classes in $H^*(BSO_{2n+1}, \mathbb{F}_2)$. Note that $c, w_2, \ldots, w_{2n+1}$ are algebraically independent over $\mathbb{F}_2$, and are homogeneous of degrees $2, 2, 3, \ldots, 2n+1$.

3.5.12 The rings $H^*(BGO_{2n,k}, \mathbb{F}_2)$. This ring is generated by the cohomology classes λ, a_{2i-1}, b_{4j}, and d_T over $\mathbb{F}_2$, where $i, j \in \{1, \ldots, n\}$ and $T \subset \{1, \ldots, n\}$ has cardinality $|T| \geq 2$. These are defined as follows. Let (E_{2n}, L_{2n}, b_{2n}) denote the universal quadratic triple on BGO_{2n}. Then $\lambda = \bar{c}_1(L_{2n})$ and $b_{4j} = \bar{c}_{2j}(E_{2n})$. Next, consider the Gysin boundary map $d : H^*(BO_{n,k}, \mathbb{F}_2) \to H^{*-1}(BGO_{n,k}, \mathbb{F}_2)$ that we introduced earlier. Then $a_{2i-1} = d(w_{2i}) \in H^{2i-1}(BGO_{n,k}, \mathbb{F}_2)$, and for any $T = \{i_1, \ldots, i_r\} \subset \{1, \ldots, n\}$, we have $d_T = d(w_{2i_1} \cdots w_{2i_r}) \in H^{2(i_1+\ldots+i_r)-1}(BGO_{n,k}, \mathbb{F}_2)$. These generators satisfy relations given explicitly in [H-N-1] Theorem 3.9.

3.5.13 Relation with Stiefel-Whitney and Chern classes. When $k = k_{sep}$, the ring homomorphisms $H^*(BGL_{n,k}, \mathbb{F}_2) \to H^*(BGO_{n,k}, \mathbb{F}_2) \to H^*(BO_{n,k}, \mathbb{F}_2)$, induced by the inclusions $O_n \hookrightarrow GO_n \hookrightarrow GL_n$, are given by the same explicit formulas in terms of the generators of $H^*(BGO_{n,k}, \mathbb{F}_2)$ which are deduced in Section 3 of [H-N-2] in the topological category for $H^*_{sing}(BGO(\mathbb{C}), \mathbb{F}_2)$.

3.5.14 With the above explicit descriptions of $H^*(BGO_{n,k}, \mathbb{F}_2)$ in the even and odd cases, all the calculations Section 8 of [H-N-2] hold in the algebraic category over such a k. These include Corollaries 8.1 and 8.2 which give formulas in terms of generators and relations for the 'even rank degenerating to odd rank' and 'odd rank degenerating to even rank' cases of the Theorem 3.5.1, illustrated with complete lists of generators of the primitive rings $PH^*(BGO_{n,k}, \mathbb{F}_2)$ and their images under ∂ in all ranks up to $n = 6$.

Remark 3.5.15 For any scheme X over $\mathbb{Z}[1/2]$ and any $y \in \Gamma(X, \mathcal{O}_X)^\times$, let $(y) \in H^1(X, \mathbb{F}_2)$ denote the Kummer class of y. If X lies over a separably closed field base field of characteristic $\neq 2$, then note that $(y)^2 = 0 \in H^2(X, \mathbb{F}_2)$, and moreover, $(-1) = 0 \in H^2(X, \mathbb{F}_2)$. These properties of the Kummer class also hold in the topological category for Kummer classes of nowhere vanishing complex valued continuous functions y. These properties, which are crucially used in [H-N-1] and [H-N-2], are not available over an arbitrary base such as $\mathbb{Z}[1/2]$. In fact all powers κ^n of the Kummer class $\kappa = (-1) \in H^1(\mathbb{Z}[1/2], \mathbb{F}_2)$ are nonzero, as may be seen by base change to $\mathbb{R}$. This is one of the reasons why the explicit determination of the rings $H^*(BGO_n, \mathbb{F}_2)$ and $PH^*(BGO_n, \mathbb{F}_2)$ in terms of generators and relations is more complicated over a general base. This question will be addressed elsewhere.

Bibliography

[Be1] Behrend, K.A. : Derived ℓ-adic categories for algebraic stacks. Mem. Amer. Math. Soc. 163 (2003), no. 774

[Be2] Behrend, K.A. : The Lefschetz trace formula for algebraic stacks. Invent. math. 112, pp. 127-149 (1993)

[B-H] Borel, A. and Hirzebruch, F. : Characteristic Classes and Homogeneous Spaces, I. Americal Journal of Mathematics, Vol. 80, No. 2 (Apr., 1958), pp. 458-538

[Bh] Bhaumik, S. : Characteristic classes for $GO(2n)$ in étale cohomology. Proc. Indian Acad. Sci. (Math. Sci.), Vol. 123, No. 2 (May 2013), pp. 225-233

[EGA-IV$_2$] Grothendieck, A. : Éléments de géométrie algébrique. IV. Étude locale des schémas et des morphismes de schémas, seconde partie. Inst. Hautes Études Sci. Publ. Math. No. 24, 1964.

[E-K-V] Esnault, H., Kahn, B., Viehweg, E. : Coverings with odd ramification and Stiefel-Whitney classes. J. Reine Angew. Math. 441 (1993), 145-188.

[Fu] Fujiwara, K. : A proof of the absolute purity conjecture (after Gabber). Algebraic geometry 2000, Azumino (Hotaka), 153-183, Adv. Stud. Pure Math., 36, Math. Soc. Japan, 2002.

[GIT] Mumford, D., Fogarty, J., Kirwan, F. : Geometric Invariant Theory. Springer-Verlag (1994).

[J] Jardine, J.F. : Universal Hasse-Witt classes. Algebraic K-theory and algebraic number theory (Honolulu 1987) 83-100, Contemp. Math. 83, Amer. Math. Soc., 1989.

[Ju] Jouanolou, J. P. : Une suite exacte de Mayer-Vietoris en K-théorie algébrique. Algebraic K-theory, I: Higher K-theories (Proc. Conf., Battelle Memorial Inst., Seattle, 1972), 293-316. Lecture Notes in Math., 341, Springer, 1973.

[H-N-1] Holla, Y.I. and Nitsure, N. : Characteristic classes for $GO(2n, \mathbb{C})$, Asian

Journal of Mathematics 5, 2001, 169-182.

[H-N-2] Holla, Y.I. and Nitsure, N. : Topology of quadric bundles. Internat. J. Math. 12 (2001), 1005-1047.

[La] Laborde, O. : Classes de Stiefel-Whitney en cohomologie étale. Colloque sur les Formes Quadratiques (Montpellier, 1975). Bull. Soc. Math. France Suppl. Mem. No. 48 (1976), 47-51.

[L-MB] Laumon, G. and Moret-Bailly, L. : *Champs algébriques.* Springer 2000.

[L-O] Laszlo, Y. and Olsson, M. : The six operations for sheaves on Artin stacks I: Finite coefficients. Publ. Math. IHÉS 107 (2008), 109-168.

[Na-Ra] Narasimhan, M.S. and Ramanan, S. : Geometry of Hecke cycles -I. *C. P. Ramanujam - a tribute*, 291-345, TIFR Studies in Mathematics 8, Springer, 1978.

[M] Milne, J : Étale Cohomology. Princeton University Press (1980).

[N-1] Nitsure, N. : Topology of conic bundles. J. London Math. Soc. (2) 35 (1987) 18-28.

[N-2] Nitsure, N. : Cohomology of desingularization of moduli space of vector bundles. Compositio Math. 69 (1989) 303-339.

[O] Olsson, M. : Sheaves on Artin stacks. J. Reine Angew. Math. 603 (2007), 55-112.

[Se] Seshadri, C.S. : Geometric reductivity over arbitrary base. Advances in Math. 26 (1977), 225-274.

[SGA 4] Artin, M., Grothendieck, A., Verdier, J.-L. (ed.) : *Théorie des topos et cohomologie étale des schm̃as*, Sem. Geom. Alg. Bois Marie 1963-64, LNM 269, 270 and 305, Springer 1972-73.

[SGA $4\frac{1}{2}$] Deligne, P. : *Cohomologie étale.* (With the collaboration of J. F. Boutot, A. Grothendieck, L. Illusie and J.-L. Verdier.) LNM 569, Springer 1977.

[Td] Toda, H. : Cohomology of Classifying Spaces. In the volume *Homotopy Theory and Related Topics* (Editor H. Toda), Advances in Pure Math. **9** (1986) 75-108.

[Tot] Totaro, B. : Chow ring of a classifying space, *Algebraic K-theory* (Seattle 1997) 249-281, Proc. Symp. in Pure Math. 67, Amer. Math. Soc., 1999.

www.ingramcontent.com/pod-product-compliance
Lightning Source LLC
LaVergne TN
LVHW082359200926
848481LV00012B/3623
* 9 7 9 8 2 3 2 3 8 7 7 7 8 *